Wishing you a lot of reading
pleasure!

Matthijs

BANKING THE CLOUD

CHRIS ZADEH & MATTHIJS ALER

WITH MARY CURRAN-HACKETT

BANKING THE CLOUD

**THE COMPREHENSIVE HISTORY OF
CLOUD BANKING AND THOSE WHO STARTED IT**

ForbesBooks

Published by ForbesBooks, Charleston, South Carolina.
Member of Advantage Media Group.

ForbesBooks is a registered trademark, and the ForbesBooks colophon is a trademark of Forbes Media, LLC.

Printed in the United States of America.

10 9 8 7 6 5 4 3 2 1

ISBN: 978-1-946633-72-9
LCCN: 2020915784

Cover design by Carly Blake.
Layout design by Megan Elger.

This publication is designed to provide accurate and authoritative information in regard to the subject matter covered. It is sold with the understanding that the publisher is not engaged in rendering legal, accounting, or other professional services. If legal advice or other expert assistance is required, the services of a competent professional person should be sought.

Advantage Media Group is proud to be a part of the Tree Neutral® program. Tree Neutral offsets the number of trees consumed in the production and printing of this book by taking proactive steps such as planting trees in direct proportion to the number of trees used to print books. To learn more about Tree Neutral, please visit **www.treeneutral.com**.

TreeNeutral

Since 1917, the Forbes mission has remained constant. Global Champions of Entrepreneurial Capitalism. ForbesBooks exists to further that aim by bringing the Stories, Passion, and Knowledge of top thought leaders to the forefront. ForbesBooks brings you The Best in Business. To be considered for publication, please visit **www.forbesbooks.com**.

If you never want to be criticized,
for goodness' sake don't do anything new.

—Jeff Bezos, founder of Amazon

CONTENTS

PART SIX: RIPPING CURRENTS (2018–2019)

THE GREAT WIDE OPEN

CHRIS ZADEH AND MATTHIJS ALER weren't looking for Nemo. Accomplished scuba divers, they were going where most couldn't venture, let alone would.

The twenty-nine-year-old men had traveled together before. They shared an intense interest in diving, travel, and adventure. They both subscribed to the "work hard, play hard" mentality. When they weren't traveling or diving, they were used to working long days. But out in the ocean they felt light-years away from the cutthroat financial world of big money, big banks, and big stakes in which they operated daily. Despite wanting a break, they were still up for a big adventure. In this case, they were going after giant manta rays off the coast of Fiji.

Manta rays can reach up to six and a half meters in width. Besides whales and sharks, they are among the largest and most powerful animals in the sea. In ancient times, mantas were feared for their size and power. In fact, the ancient Peruvian Mochicas worshipped mantas, and many of the Mochicas' artworks depict them. Over the years, the mantas have come to signify one's journey—the great adventure into the unknown. Forward swimming (they physically can't go backward) and intelligent (mantas have one of the largest brain-to-body ratios among sea creatures), mantas know only one direction: straight ahead. Agile, powerful, and forward thinking, mantas also signified what they were going after in both work and play—the big fish.

Nothing less would do.

In fact, whatever Chris and Matthijs did, they gave it their all. Competitive, ambitious, and ruthless, the two men weren't afraid to go where most others dared not. That's not to say they were merely thrill seekers. Sure, they loved the risk. But they weren't the types to be sloppy or reckless. Educated, knowledgeable, and cautious, the men never underestimated the dangers that lay below—seen or unseen. They did their homework and came prepared. They took their safety and the safety of others, not to mention the safety of the fish they were diving with, very seriously. They always took the necessary precautions. They recognized that there is no redundancy in scuba diving—no extra oxygen tanks, regulators, or masks. Like the world of core banking systems they operated in, there was absolutely no margin for error.

Sitting on the edge of the boat in September 2005 in their scuba gear and wetsuits, the friends both looked out at the great wide open—large clouds casting shadows over the deep blue ocean. It was beautiful. It was vast. And there was so much more below the surface than the eye could see. So much to explore. So many unknowns.

They were ready to dive in.

Though the best of friends by then, Chris and Matthijs made an unlikely duo at first. Just looking at them speaks volumes. Matthijs looks quintessentially Dutch—the poster boy for Holland. He is tall and lean with a full head of perfectly coiffed blond hair. He wears button-down shirts, trendy glasses, and sensible shoes for cycling to work. He often looks serious—perhaps even intimidating at first. But his polished and aloof look belies his wicked sense of humor and easy smile. Even tempered, thoughtful, and measured, Matthijs has an outward calm that often masks the internal wheels that are always turning.

Chris is the yang to Matthijs's yin. Though Chris's mother is 100 percent Dutch, his father is both Russian and Persian. He has dark piercing eyes and thick dark

Chris is the yang to Matthijs's yin.

hair that spikes out on the sides like the small tufts of a great horned owl. He is average height but muscular and solid. His arms, which are covered in Polynesian-styled tattoos of sea creatures that are deeply significant to him, bulge through his sweaters. An avid martial arts enthusiast, Chris isn't one to shy away from a fight or letting people know what he thinks. Energetic, passionate, and outspoken, he uses his whole body when he speaks. One might say that when they are together, they are a whole person. Chris says that together they are better than two—they have the power of three men.

The two first met in 2001 while working for the start-up BinckBank. BinckBank was the first in Europe to offer online stock trading at rock-bottom prices. Chris had joined the company first in early 2000. The founder of BinckBank, Kalo Bagijn, took him under his wing and asked Chris, who already had experience with online trading, to help him with his start-up. Luckily for Chris he was an unlikely candidate for most other jobs. He jokingly likes to tell of how he was turned down to work as a telephone operator and was fired as a waiter after one day. He would be the first to admit that it was all for the best. Most who know him would agree that Chris would never have been suitable for jobs where he had to take orders from anyone—except Kalo. A hard worker nonetheless, Chris found out early he was proficient in other areas, namely the area between business and technology. In fact, his first job in banking was at ING as a stockbroker at just nineteen years old. With this job he paid his way through college. He eventually left university without graduating when Kalo asked him to become the tech guy at BinckBank. Always on the vanguard of what was happening in technology and banking, Chris had a curiosity about finance, technology, and how they converge that was insatiable. Kalo, his boss and mentor, liked that in him. He also liked Chris's unapologetic drive and unwavering ambition. At just twenty-three years old,

Chris began building online applications to run the bank and worked his way up to the position of managing director of the Netherlands as Kalo's right-hand man. In just nine years, by 2008 Chris had learned how retail banks worked from the inside out. He knew them so well he could build one of his own.

When Matthijs showed up for a job interview with Chris in 2001, Matthijs was less than impressed by his young "boss." The feeling was mutual. Chris begrudgingly agreed to meet Matthijs, who Chris felt was just another college frat boy looking for an easy internship to pad his résumé. Chris, a college dropout himself, didn't have much patience for people who flaunted their college degrees and who didn't have a willingness to roll their sleeves up and work hard. Matthijs, on the other hand, didn't much appreciate people who were quick to judge.

The two were off to an awkward start.

When Matthijs walked into BinckBank for the first time, no one went out of their way to make him feel welcome. When he finally caught someone's attention and found his way to Chris's office, he found Chris with his feet up on his desk. *A power move*, Matthijs thought to himself. *He's showing me who's the boss.* In reality, Chris had just injured his knee in a scooter accident and needed to keep it elevated. Regardless, Matthijs was irritated. Not one to miss the details, Matthijs noticed Chris had crutches nearby and could have easily walked the fifteen meters to the door to greet him. Admittedly, Chris recalls, he wasn't all that eager to meet the new hire. In the end, Chris had no choice but to hire Matthijs. Kalo was impressed by the research Matthijs was conducting for his thesis. More importantly, he was excited that Matthijs was working on this thesis with a top professor at Erasmus University in Rotterdam in the faculty of business administration. Kalo thought this might lead to good connections down the road. Matthijs was, to put it bluntly, a means to an end. Kalo more or less told Chris that he didn't care if Chris liked Matthijs. They didn't need Matthijs in the long run. They needed those whom his professors knew.

If Chris was going to have someone on his team he didn't like or approve of, he wasn't going to make it easy on the guy. The first thing Matthijs recalls Chris saying to him was, "I'm on the phone and talking about confidential stuff. If you go back to Rotterdam and talk to your frat buddies, and I hear about it, I'll throw you out the window."

Not one to be intimidated by anyone, Matthijs simply said, "Okay" and dismissed the comment as nothing more than it was—a guy reminding him how important his job was at a bank and how important his client's security was. *Got it*, Matthijs thought.

Their stalemate didn't last. Chris quickly realized he'd underestimated the quiet, reserved scholar his boss had made him hire. Within a few days, Chris learned what most people do about Matthijs once they spend some time with him: he has a quick wit and a wicked sense of humor. In short order they made each other laugh. Within a few days of working with Matthijs, Chris went home to his girlfriend and said, "You know, I have this new dude, and I spend half the day pissing my pants because he makes me laugh so hard." Both men believe it's a great strength not to take oneself too seriously. And both men are adept at it. They love nothing more than making funny jokes about themselves, each other, and of course other people. Matthijs's humor is especially acute. He thinks so fast and can make a comic remark within seconds—leaving most people's mouths still ajar in astonishment as he nonchalantly moves on to another topic.

Though laughter was the salve that healed their original displeasure with each other, what really bonded the two was the work.

They would work until very late and have dinner in the office almost every day. No matter how complex the issue, the two came up with good ways to solve it. Soon both realized that they worked better together than alone. Bouncing ideas off each other was an integral part of their process. Chris was impressed by Matthijs's ability to see through complex issues and come up with solutions. And Matthijs was impressed with the way Chris could take a derailed project from someone else and put it back on the rails. Matthijs also took note of Chris's leadership style. Chris was known throughout BinckBank, to put it euphemistically, as *unfriendly*. But, in the end, it was results that counted to Matthijs. The results were always outstanding and always in the interest of the company—not Chris's ego. Matthijs liked that about him. He also admired how everybody seemed to understand that they needed to comply with the way Chris said the project should be executed. He had a way about him that would align everyone to one singular vision. Matthijs recalls at one point there was a big IT project at BinckBank that entailed the migration of a core banking system from A to B. The entire platform was being upgraded. It was a complete nightmare under the direction of the IT department head. Kalo asked Chris if he could step in. Chris took an inventory of all the issues and prioritized them. Then, systematically, Chris and his team started tackling the issues one by one. Important ones first, less important ones later. And within a matter of weeks, the project was wrapped up and a total success. Matthijs remembers watching how Chris led the team and was utterly impressed by what his friend had easily accomplished.

Matthijs mostly admired this particular quality in Chris: his absolute confidence and 100 percent conviction. Chris never has a shred of doubt in his mind about his decisions. Matthijs, on the other hand, may have a very strong opinion about what the way forward is in any given situation, but if he gets feedback or new information, he could be led to a slightly different perspective or an entirely different view of the way something should be done. Chris doesn't know the word *doubt* and isn't easily persuaded. With Chris's conviction and Matthijs's willingness to execute, they made quite a team, both at work and at play. If there was anyone Chris trusted with his life when diving into a vast ocean or a new project, it was Matthijs. And Matthijs felt the same.

> **With Chris's conviction and Matthijs's willingness to execute, they made quite a team, both at work and at play.**

Seven years later, the men were sitting on a different precipice than the edge of a boat in the middle of the Pacific Ocean. In 2012, instead of leading a team at BinckBank, they were at the helm of the company Chris founded, Ohpen. If they could pull off what they were about to do, they would be the first company in the world to build a core banking engine of a bank—from scratch—and launch that bank in the *cloud*. Talk about the great wide open.

Though vastly misunderstood, the cloud is not that difficult to explain. Simply put, cloud computing is the on-demand delivery of technology resources (including servers, storage, databases, networking, software, analytics, and intelligence) over the internet with pay-as-you-go pricing. Instead of buying, owning, and maintaining physical data centers and servers, you can access technology services on an as-needed basis from a cloud provider.

Most people use the cloud every day to stream videos, play video games, access web mail and social media, or operate office-productivity software. Instead of accessing data or files from a personal computer or from a network of servers on-premises, when in the cloud a person is able to access them online from any internet-capable device—so the information will be available anywhere, anytime.

In 2008, most people were just hearing about the cloud, and no one, let alone banks, thought of using the cloud instead of building their own IT systems. Unlimited computing power without the hassle of buying, installing, configuring, patching, or upgrading it— the dream of any innovative IT boss *if he/she "gets" it.*

Up until Ohpen, no bank had operated applications in the cloud before let alone stored, used, and processed production data in the cloud. Rather, every single bank in the world built its own platform, what is known today as a legacy system. A bank would build a data center (or rent it if they were not big enough to build their own). Then they purchased or built software and hired developers, application managers, and systems engineers to tie it all together into one system to further build, change, patch, and maintain all these systems. They managed all their software and stored all their company's information in giant computers in this so-called data center in one location. Each bank essentially reinvented the wheel every time they set up their banking systems. It is like every bank would build their own factory. Even Apple has its own factory. Why would fifty thousand banks all have their own factories if they all do the same thing? And what are the results of all this? Massive issues, massive amounts of time wasted, massive workforces required, and massive expenses.

No one knew this better than Chris. He had spent years working at BinckBank and knew how to build a bank from scratch. Sitting in an office full of developers and IT guys, Chris knew all the building elements required. He personally understood the frustration his IT guys felt trying to build a platform on a bank's legacy system. If someone wasn't paying attention to where they were walking, they could trip on a cord and the entire server would go down. If upgrades or fixes needed to be made, they would have to patch it and rewire it. At the end of several years, the server wires ended up looking like a bowl of spaghetti. In addition, if someone wanted to make a change to the program, it would require loads of red tape. It could take months to get anything done. In a word: it was a *mess*.

After eight years of working tirelessly for BinckBank, in 2008 Chris's boss, Kalo, resigned, and Chris decided to as well. Kalo told Chris he was the only guy who had the skills to be his successor, but in the same sentence he told him he would not become the CEO. Too many people were against it. Nevertheless, Chris made a plan to reprogram the whole core banking system of BinckBank. After eight years, nothing worked anymore, and BinckBank had outages and failing systems every day. Chris said, "If we don't do this now, in ten years the billion-dollar valuation on the bank will be gone." The new executive board had no interest in Chris's plan. So he left.

Instead of going to another bank and having another boss to report to, Chris decided to take a year off work and travel the world with his then-girlfriend (now wife), Myrthe. While on his travels, he used the time to do some research on best practices in the tech industry. This led him to Silicon Valley, where he visited different companies, read the local newspapers, and talked to a variety of industry people. It was in the valley that he stumbled upon the cloud. As soon as he saw it, he immediately thought: *Unlimited computing power? Guaranteed uptime? Autoscaling? Only paying for what you use? This is the biggest revolution since computers were invented!* Not one to let things go or leave an idea to wither on the vine, Chris began to do a deep dive into researching cloud-based servers and started to investigate Amazon Web Services (AWS). As one of the largest cloud-based servers in the world, its functionality, scope, capabilities, and safety measures were astounding. It soon became clear to Chris that AWS had what he called the "important things," namely a better working IT infrastructure than most banks could build themselves. It was about this time that Chris had an epiphany: *I am on to something. I know how to build a bank. I know what this cloud thing can do. If I could gather the right team around me, I could build core banking software and put it in the cloud!* He could then sell this software as a service (SaaS) to banks all over the world. They could get rid of their hardware and of the core bank applications, spaghetti wiring, about 80 percent of their IT operating costs, and have a safe, agile, and perfectly running banking software that worked every time, everywhere. They could just rent the platform.

Chris thought: *The banks could focus on their customers and leave the rest to us.* In his head it made total sense: *IT. JUST. WORKS.*

Rested from his sabbatical and energized by this new idea, Chris knew what he had to do. And nothing was going to stop him. He was all in. And he knew just the guys to do it—all his friends from BinckBank, Bas Wouwenaar, Ilco van Bolhuis, Erik Drijkoningen, and of course Matthijs.

The first guy on the list for Chris to call was his ride-or-die friend, Matthijs.

In Chris's mind, this was a no-brainer. They worked so well together. Besides, Matthijs owed him one. Not long before leaving for his trip around the world, Chris arranged for a promotion for Matthijs to set up the new BinckBank office in Paris, France, and be its CEO. It was an amazing opportunity for Matthijs. But, Chris thought, the opportunity he was presenting to Matthijs was even better. Here was Matthijs's chance to be a leader with him—and not just any leader but an owner and founder of a new company that would be the first of its kind. Together they would build the world's first cloud-native core banking system. The sky was the limit. He was going to service *banks*. Talk about big fish! This would be their biggest adventure yet.

So Chris was stunned when he heard Matthijs's response to his invitation: "No."

Simple and to the point, Matthijs just couldn't do it. He lived in Paris, had just gotten married, had a baby on the way, and had just launched Binck.fr. It was simply too risky. He had a good job that wasn't finished and a life he enjoyed in Paris. Chris tried to persuade him. But Matthijs stood firm. Chris didn't do waiting. He was ready to go. He was also devastated and hurt. He simply couldn't fathom how Matthijs could say no to him after all he had done for him over the years. He'd given Matthijs his first job. He'd given him a promotion. He was the reason Matthijs was running the Paris office for BinckBank. But Chris was undeterred. Yes, it would be more challenging to pull off without Matthijs, but he could do it.

The second person on his list to call was an old friend he'd known since they both were sixteen years old and an IT octopus from BinckBank, Bas, whom he knew he could count on to execute.

After Bas, Chris called Erik, who was hands down the best product developer in the industry. Erik had recently left BinckBank to join a consulting firm in Germany. Before Chris could get the question out, Erik's response was "Yes."

"You don't even know what I am going to ask," Chris said.

"I don't need to. I'm in," Erik declared. He knew Chris well enough to know that if he was calling him in Germany, it was because he had a job for him to do. Feeling unfulfilled and ready to move on from consulting, Erik was ready for a new challenge, and there was no one he would want to work with more than Chris. Because with Chris, he knew it would get done.

When Erik, Bas, and Chris met to discuss who they would hire as coders, they unanimously decided the first one would be Ilco, whom they all knew from BinckBank. In their minds, there was simply no better hardworking coder out there. If you could speak it or imagine it, Ilco could code it. When Ilco dreamed, the team believed, it was in zeros and ones. What Chris liked about Ilco was the standard by which Ilco believed a project or product or feature was working: *when it was working in a live environment*. In other words, *it works when it works*. And, Chris adds, "It's on him to get it to work." For two years, the team worked around the clock to create the platform while Chris tried to sell it. Eventually, Matthijs couldn't resist the pull of working alongside his friends and left BinckBank in France to join the Ohpen team in Amsterdam officially in 2011.

By then there was no time to rest; Matthijs and Chris had the fight of their lives ahead of them.

If they were going to catch this big fish, their own proverbial manta ray—a forward-thinking and intelligent bank with an eye on the future—they had to overcome the impossible: they had to prove to the financial industry that their platform was working and ready to use. They had to prove it was reliable, adaptable, and compliant. They had to do the nearly impossible, too: they had to get the Dutch National Bank (De Nederlandsche Bank, or DNB) to approve of a bank, hence its customers' financial and processing real-time data and information, to sit on the servers of an American-based company, *an online bookstore*. If they received this approval, they would be the first company of its kind to be granted such permission in the world. And once all the contracts were signed, all the safety protocols, regulations, and compliance measures were met, they had to migrate all of a bank's data from its original on-premises servers to the cloud—successfully. *And* they had to do it with one of the smallest banking IT teams in the world. No other bank could do more with less than what they were working with. If they succeeded, they would make history and change the financial industry forever. If they failed, Chris would lose his €3 million investment, which was dwindling by the day. His team had sacrificed everything. For four solid years they'd worked up to eighteen hours a day, seven days a week. The men who had children at home had not seen them during waking hours in all that time. They gave it their all because they believed in what Chris was trying to achieve; moreover, they believed in their own unique and exceptional talents that could help Chris pull off the impossible.

On Easter Sunday, March 31, 2013, four years after Chris's grand idea and two years after Matthijs finally agreed to join the team, the two stood on yet another precipice together. They looked over a team of exhausted but excited developers and coders who were all sitting at a circular boardroom table, which had become a de facto NASA-like mission control center. If the team pulled this off—if they successfully migrated a major bank's data onto the platform they'd built—they would be the first company in the world to put a bank in the cloud.

The anticipation was palpable. It was Easter Sunday. Everyone on the team had left their families and given up their holiday to see this through. This was it. They were all in.

Just like in their diving days, Chris and Matthijs were prepared for every scenario. They had practiced for this. They had done their homework. They had covered all the safety protocols. They were compliant with all the regulations. For nearly seven straight days, their whole team had never left the office. They'd worked around the clock. If they succeeded, it would be the beginning of a new era of banking. If they failed, there was no turning back and all the hard work would have been in vain. They could pack their bags and seek new employment. The big fish would forever elude them. This would be the end of Ohpen and everything they had worked for and achieved up until that moment. Everyone in the room was counting on it to work.

They all waited and watched as the information migrated—billions of euros' worth of their client's assets. Their first client, the Dutch bank Robeco, whose offices were in Rotterdam, was counting on them. They too were taking a huge but calculated risk on this small start-up with a big idea.

Matthijs and Chris didn't flinch. They didn't panic. It was all systems go. They would bet their life on its working. *IT. JUST. WORKS.* was a constant refrain.

They had plunged to the depths of the ocean. They had swum alongside the giant manta rays and watched as they slowly and gracefully moved forward against powerful currents that resisted them. They had seen their power and their willingness to move forward despite all obstacles. They had taken the plunge into the unknown before. They had seen their risks pay off. They had seen their hard work and preparation come to fruition.

This would be no different.

As the data uploaded before their eyes, Ilco, Bas, Erik, Matthijs, Chris, and all the members of the team that had expanded over the past four years exchanged glances. This was it.

They were diving into the great wide *Ohpen*.

PART ONE

PREPARATION

CHAPTER 1

WHERE IT ALL BEGAN

SITTING IN THEIR SLEEK OFFICE on the fourth floor of an art deco–era building located in the heart of Amsterdam, Chris and Matthijs have one of the best views of the city. Everything about their location is quintessentially Dutch. A sea of bicycles is parked outside. A modern tram runs down the street past historic buildings with their iconic stepped gables. The area is rife with history. Across the street, a large pillar memorializes the site of the "Miracle of Amsterdam," a legend that dates back to 1345. Just meters away from their office is Dam Square, site of the Royal Palace of the Netherlands. And not far from there, centuries ago Amsterdam's first commodities exchange was built in 1608 at the dawn of the Dutch Golden Age. Though history books say it ended centuries ago, the golden age is alive and well in the hearts of the Dutch, who still celebrate their art, government, trade, and financial prowess every bit today as they did then. And perhaps there are no two prouder men to be part of such a rich history than Chris and Matthijs, who feel just as proud to be part of the creation of its future or, rather, what they see as the dawn of a new golden age—a technological age in finance.

Below the large windows, the two have desks that face each other, making it easy for them to discuss whatever matters are at hand. The men look relaxed, seemingly unfazed by the frequent calls that come in and never seem to cease. When an employee arrives at their office, Chris, who faces the glass door, waves them in and gestures for them to sit on one of their two large black leather couches. However, in order for anyone to enter, they have to pass a long boardroom table flanked by silver paintings that represent the Chinese elements (water, wood, fire, earth, and metal). Then, before entering, they must pass replicas of terra-cotta warriors that sit on guard outside their glass doors, which serve as a constant dual reminder of what the men value: security and the warrior mentality. Finally, in order to enter any room in the building, even their office, they have to press their pointer finger on a sensor pad. No one (visitor or employee) is exempt.

Sitting across from the men, one can easily take the entire experience for granted—the beautiful Zen-like office, the sense of confidence the two men have in themselves and the company they now run, and their optimistic future. But their individual journeys to the C-suite and to making financial history as the creators of the first cloud-based core banking engine in the world were hardly a given. Their journey was an arduous one.

Perhaps the reason Chris values security so much today—one might say he is zealous about it—is that he himself had none growing up.

"I had no safety net. No safe house. No safe place to land," he says, explaining his childhood to Matthijs one spring afternoon over lunch at their boardroom table, where the two often eat meals together. But Chris is also quick to acknowledge, "Even though I didn't have much of a family, I always knew there were other places in the world that had it a lot tougher."

This wise perspective came to him after learning about (and making peace with) how his own father made his mom, brother, and himself flee their Parisian home when he was just four years old. When he was a teenager, Chris learned his own father came from an abusive past. His grandfather, who came from Persia (now Iran), was a violent man. He beat his wife and his seven kids regularly. Chris recalled his own father telling him that he used to have to pick up his mom (Chris's grandmother) from the hospital many times. Chris said his father, aunts, and uncles still have visible scars from the brutal beatings.

Chris's mother didn't have it any easier. Though not beaten, she suffered from abandonment issues. She was what is known as a "farm child." During World War II, people were sent to farm basements to hide and stay safe. Her biological parents met and conceived her there, but they didn't want to raise her. The family who had given the couple shelter agreed to raise the young couple's child, Chris's mother. Her start in life is something she never fully recovered from, and so she spent her young adulthood feeling lost and unmoored.

Needless to say, Chris's parents had a lot of psychological damage to overcome when they met in France and started a family together. And like a lot of couples with two small children, the responsibilities and pressures of marriage and raising a family became too much. After Chris's mother left his father, he became an alcoholic and refused to pay alimony. Chris's mother moved to the Netherlands seeking security. Those were some of the most difficult years of Chris's life. His mother was alone, raising two small children and working constantly to support them. She struggled to not only put food on the table but also keep a roof over their heads. Chris once overheard his mother begging his father for money over the phone late at night after the landlord said to her, "If you don't pay the bills now, you will be on the streets with your children."

These overheard conversations left an indelible mark on young Chris. He vowed from the earliest age never to find himself in a similar situation—fearing where his next meal was coming from and wondering whether or not he would have a place to live.

As if being poor and not feeling secure at home wasn't enough of a struggle, Chris faced even more instability and trauma when he went to school.

"I didn't speak the language. I was like an alien," he says about starting in Dutch schools. He knew no one and was overwhelmed by loneliness. Everyone else seemed to know each other and get along. From day one, he felt like an outsider.

He found his stride outside and in playing sports, and he hoped to someday become an elite athlete. But no matter what he did, when he returned from school or sports at the end of the day, he came home to a bleak and sorrow-filled home. "I knew my mom loved us, but she just had sorrow. Sorrow to pay the bills. Sorrow not having someone to be with … but as a kid, sometimes you just want someone to give you a hug. You want someone to say it's going to be okay. I didn't have that. She didn't have that."

His mother's sadness was palpable at home, and he longed for an escape. He spent most of his time as a teenager with his peers. Left to his own devices, he had to figure out life by himself without much help. He became an adept problem solver and was resourceful, and he found work at just fifteen years old. His father never came to his sports matches, and he didn't even show up for his high school graduation. When Chris tore a number of ligaments in his knee, his dreams of becoming an elite athlete were immediately dashed. Like many of his peers, Chris turned to numbing his pain—physical and mental—with marijuana. He smoked three grams a day for five straight years, and it progressed to ecstasy and cocaine on the weekends when out partying with friends. After a drug-induced anxiety attack at age twenty, Chris never touched another drug again. He went cold turkey and never looked back. "I haven't touched anything in twenty-three years, and I'll never touch it. Something just clicked in me. I wanted my mother to be proud of me. I didn't want her to have struggled for nothing. I didn't want to struggle like my own mother or father, so I decided to give it my all, always and always."

Chris wanted to be successful at something—actually, anything he put his hands on—so he applied the same sense of self-discipline and resolve that he used to overcome addiction to his newfound passion: weightlifting. After some time, however, he began to experience back trouble, and a friend encouraged him to try yoga to relieve the pain. What began as practice to alleviate physical pain became a way of life. Step by step he started to incorporate meditation into his practice as well. It wasn't long before he began sleeping better at night, and he began feeling a sense of peace, acceptance, and forgiveness. The more he did it, the more his own journey in life began to make sense. "It really helped me to leave the past in the past and not become bitter." The more Chris reflected on his childhood, the more grateful he became. "It made me who I am. It gave me determination and perseverance." Determination, perseverance, self-discipline, adaptability, and passion were the values that helped Chris climb out of the ruins of his childhood. Proud of these values and never wanting to forget them, Chris had them tattooed symbolically on his body in the form of Polynesian-styled sea creatures that represent each

> Determination, perseverance, self-discipline, adaptability, and passion were the values that helped Chris climb out of the ruins of his childhood.

virtue. One of the most prominent is the hard-bodied, soft-hearted crab, which is the symbol of perseverance.

An avid reader, Chris spent his young adult years reading motivational books to stay inspired and push himself. He credits books by Tony Robbins for motivating him as a young professional. He didn't have any mentors as a boy or young man, so he also looked up to icons like Richard Branson, Steve Jobs, and Arnold Schwarzenegger, who was Mr. Olympia five years in a row. But it wasn't because he was a weightlifter that Chris admired him. Schwarzenegger was the ultimate underdog. "He wasn't born the Terminator," Chris explained. "Nobody ever believed in him. When he arrived in Hollywood, they said he was too big. They said he might get a movie role as a bouncer somewhere, because he had a stupid name, a stupid accent, and because he couldn't act. But he was determined. *He* believed he was ready for a leading role. He worked his ass off to get there. He visualized his goal. Nothing stopped him. I liked that mentality. I always loved his story. Defying the odds, working your ass off, and having a vision: that's the formula to anyone's success." One could argue that it's Chris's own as well as his company's. But he adds, "Schwarzenegger didn't just defy the odds—he exceeded everyone's expectations, his own included. He not only became an actor; he became a top-grossing one. Then he became the governor."

From a young age, Chris had a soft spot for anyone like Schwarzenegger—people who liked to defy odds and prove others wrong. In fact, he's the first to admit it when someone blows his mind or surprises him. And one of those people who did just that was Matthijs, who came from an entirely different pedigree. Nevertheless, Chris knew a competitive underdog when he saw one.

Though born only months apart, the two had childhoods that were light-years away from each other. Matthijs's father had a steady job—he was a government employee and eventually a criminal judge. His mother, a nurse and then a physical therapist, worked out of their home. "My mother was supportive in every way. She was always there for me—and always at home. When I would come home from school, she would be sitting at the table with tea and cookies." Matthijs laughs at the happy memory. "She was—and still is—the perfect mother." Though she worked at home, she adjusted her schedule around Matthijs's and his siblings' schedules so she could help them with homework and cook them dinner.

His parents were heavily invested in their children's education. Matthijs and his siblings attended good schools in The Hague and then in Scheveningen, Eerste VCL, where the school motto was "I think for myself." It had a rigorous curriculum and demanded excellence from its students. Matthijs's father demanded the same from his son as well, a pressure that Matthijs felt from a young age. Matthijs's father had studied not only law but also mathematics and physical science. As a young teenager, like most boys Matthijs struggled to understand his father. Not one to give his son a pass or an easy way out, Matthijs's father liked to make Matthijs think through problems on his own. "I would do my homework and prepare for an exam, and instead of telling me how to do it, he made me go jump through all these hoops. As a kid, I just wanted to finish my homework and get to the end result faster. But he would ask me questions. There would be uncomfortable silence. And eventually he would pull it out of me." As an adult, Matthijs now sees the genius in this methodology, and he credits his father's approach to his own ability to think through problems and his relentless search for answers.

Modest about his own abilities, Matthijs refers to himself as a jack-of-all-trades instead of having a specific talent. "I couldn't name one thing, but I think as a character trait, I want to know how things work, and I won't stop until I do." This investigative spirit, he'd be the first to point out, is not a talent. "It's just *in* me. Curiosity is in me. I know how to find the information quickly—how to get to what I need to know." When he was a boy, the tools he had at his disposal were books. He was (and still is) a voracious reader. And now he'll scour websites for the information he needs. And then, like he did as a boy, he will get out some pens and paper (or an Excel spreadsheet) and do his proverbial homework to get the answers he is looking for.

Though by no means the smartest student in his class—he struggled at exam time with the German language—he could pull anything off by both studying intensely and by sheer force of will. His father, out of concern that Matthijs would fail his German-language exam and not graduate, once sent him to stay with his grandparents to study. His grandfather happened to be the preeminent expert on the German poet and writer Johann Wolfgang von Goethe. At one point his grandfather even served as the president of the Goethe Institute in Germany. He was also an aesthetics professor. His grandfather, who lived a stone's throw away from the Rijksmuseum, would take young Matthijs and his siblings on what amounted to private professional tours of the museum. His grandfather was a font of information and knew all sorts of amazing facts about all the artwork and artifacts. So when it came time to prepare for his German exam, Matthijs's father knew where to send him.

Living in The Hague at the time, Matthijs had to get on the train and head to Amsterdam by himself. When he arrived, his grandfather was waiting for him and had set up his study for him, where he spent the entire Saturday. "I remember it like it was yesterday. He was wearing a long velvet robe, and I was sitting at a table with my books, pens, and paper, and he walked around the table and asked me all these questions, and then he would tell me all about the language." Later that night, after dinner with his grandparents, Matthijs joined them for a night at the opera. Matthijs was the youngest person in the room by thirty or forty years, but it was, he recalls, an amazing experience. It was the first time the *Dreigroschenoper*, an opera by Bertolt Brecht, was being performed and celebrated. It had been so long, in fact, that people were weeping and crying because of the memories coming alive again. Later that evening, his grandparents kept him up drinking cherry liquor and listening to opera until three in the morning. But three hours later, his grandmother was standing over him, telling him to wake up—he had more studying to do. So back Matthijs went to the table.

All this work paid off. He ultimately passed his exam and graduated. Moreover, he learned many valuable lessons—he could work hard and play hard and, no matter what, always leave room for art and culture. To this day, Matthijs is a bit of a Renaissance man, and he carries with him his grandfather's encyclopedic knowledge of art, culture, and languages. He speaks and reads in three languages— English, French, and Dutch—but can quickly pick up and speak other languages as necessary. He doesn't consider it a special skill, though. "It's more of a party trick," he says. An agile conversationalist, he demurs taking credit for this as well. "That's all my mother," he explains. "I get all my social skills from her." Whether talking about finance, banking, literature, art, food, wine, current events, movies, television, or random bits of trivia he has picked up along the way during his research binges, he can move deftly from topic to topic, pulling from a deep reservoir of knowledge he has accumulated from a lifetime of deep dives into the subjects. "What I lack in natural talent, I make up for with hard work and determination."

Like Chris, Matthijs also found release and joy in sports. It also was a positive channel for all of his competitive energy and drive. He wasn't the most talented field hockey player, but like in his studies, what he lacked in talent he made up for with grit. His peers could always rely on him. This work ethic and competitive nature served him well off the field too.

Competitive, diligent, hardworking, and quick-thinking, Matthijs is a perfect match for Chris. The two often spar with one another and talk quickly, sometimes finishing each other's sentences or completing the other's thought. Though coming from two vastly different childhoods, the two have similar values and appreciate where each other comes from. Matthijs understands and appreciates why Chris values safety and a sense of security. He too values it in the same way. Though both men came at it differently—Chris lacked it, and Matthijs had it—both nevertheless recognize its vital importance. "Without feeling safe and secure, nothing else here is possible," Chris often says. Both also have incredible drive and competitive spirit. While Chris looked to people like Schwarzenegger and inspiring leadership books and Matthijs to his own father, grandfather, and works of literature, both men know the value of looking outward for inspiration. And both have a sense of gratitude for the sacrifices their hardworking parents (Chris's mom and Matthijs's mom and dad) made for them. Though being so brilliant and successful, neither feels arrogant or has an outsized ego. In fact, both are quick to say that neither could do what they do alone—both crediting their wives, Myrthe and Siena, respectively, as well as each other for their own success. This sense of deep gratitude, not to mention that there is always work to be done and that others are counting on them, has pushed both men to perform and excel—not just for their families and employees but for their clients and their clients' customers as well. At the end of the day, both men know what is on the line. They know that bank customers are people just like them, just like Chris's mom and Matthijs's parents—hardworking people trying to provide the best for their families and loved ones. Neither Chris nor Matthijs has ever lost sight of that, because they have never lost sight of where they came from.

CHAPTER 2

OUR FIRST UNICORN EXPERIENCE

"THERE IS NO GREATER TEACHER THAN EXPERIENCE," Chris says, speaking fondly of his own life, and he believes that it's true for all. Chris often advises young people who lack direction (whether in their lives or in their careers) to start at the library, though. "There may be a hundred topics to choose from, but I ask them to notice where their eyes go as they walk among the books. Wherever their eyes land—that's the answer they are looking for."

Chris, of course, is drawn to books about diving, the world's oceans, and sea life. His boardroom credenza displays a vast collection of ocean and diving photography books. But in the end, the books will only take you so far, he acknowledges. Books are great at showing someone a passion or interest, he argues, but reading books about diving off the coast of the Polynesian islands or building a bank are no substitute for diving into the real deal and experiencing them for yourself.

Chris is a firm believer that anyone can go far when they've tapped into their passion and interest, whether it's martial arts, diving, finance, or technology. But it's really experience that will take people and their passions all the way. The first time Chris had an inkling of what he was passionate about and therefore wanted to know more about was in 1995. At just nineteen years old, Chris went backpacking in Australia and visited what was then called an internet shop. The World Wide Web was in its infancy, and most people didn't have desktop computers connected to the internet in their homes, let alone mobile phones. The first time he sat in front of a computer and read an online newspaper, his curiosity was piqued. He couldn't believe what he was seeing. "I can see the newspaper from the US in Australia in real time!" He started wondering, "What's behind it? How does it work? What happens? How is it possible?" This insatiable curiosity informed him of his newfound passion: the internet.

Eventually he purchased many books on computers and computer companies. As soon as he returned home, he set out looking for opportunities to gain as much experience as possible.

While in law school Chris started working at a bank, where he first saw and experienced computing and banking processes up close. It was there that Chris found out he had a knack for it. He was overwhelmed by the possibilities of what could be improved, but he was shocked by how inefficient and costly certain processes were. "It was like becoming unplugged from the matrix. *I could just see it all.* I just knew how to make processes work better. It was like second nature," Chris says. "I was constantly telling my boss how to improve our processes, but he said I should shut the fuck up and go back to work."

It was at this job that he also began to teach himself about the stock market. One day, while sitting at his computer, he had another startling revelation. "I took a look around at all these stockbrokers on the phone making trades, and it occurred to me: *These guys won't exist much longer. Computers will do what they do. I have to leave here,*" Chris recalls.

In the beginning of the year 2000, Chris met an entrepreneur, the founder of BinckBank, Kalo Bagijn, who approached Chris with an opportunity. Chris met Kalo via his investor. "When I left ING, I wanted to go to New York and see how the biggest stock exchange in the world works," Chris recalls. "So I wrote to Dutch stockbroking firms with offices in New York, and one of them was the one that financed Kalo's idea." They took a look at Chris's résumé and saw that he had experience with the internet and coding websites, and they sent his résumé to Kalo. Soon after that, Kalo called Chris. "That's when he changed my life." Kalo explained to Chris his big idea—that anyone with a computer, no matter where they are in the world—

whether it's Australia, Shanghai, New York, or Amsterdam—could log in and make a trade in real time at rock-bottom prices. Chris recalls Kalo saying, "We're going to be the low-cost carrier of buying and selling stocks online. Do you want to build that system for me?" Chris's response was simply, "Fuck, yeah." Though he was in his third year of law school and only needed one more year to complete his degree, Chris quit university and joined Kalo. Chris and Kalo quickly hit it off, and Chris looked to Kalo as his mentor. Chris's next step was to call his longtime friend Bas Wouwenaar, whom he had met and hung out with in Amsterdam as a teenager, and invited him to join him at BinckBank. "Come with me," Chris said.

Like Chris, higher education wasn't for Bas, and he had dropped out of school as well. He was working at Getronics, a major IT company in the Netherlands, when he got Chris's call. Bas's response was friendly but to the point: "No!" He was all too happy with his current employer—he had a good salary, a company car, a mobile phone, and a laptop. But Chris was persuasive. Chris asked him, "Are you going to drive around in your ugly Getronics car all your life and be a gray mouse, or shall we make something amazing?" Bas knew Chris was right, accepted the "ego-push," and quit his job to join Binck. "I was employee number ten. And there was nothing at Binck IT-wise," Bas recalls with a warm smile, his bright-blue eyes revealing a sense of satisfaction knowing he was there from the beginning. When Bas arrived at the new Binck office, there were even some servers still in plastic. For Bas, it wasn't just exciting to be a part of something new; it was his chance to gain valuable experience and—more importantly—to do so with his friend.

It was there that both Chris and Bas met Erik Drijkoningen. Chris interviewed Erik for his job at Binck, and in that interview Chris clicked with Erik. Chris commiserated with Erik about not liking law school, and Erik explained he'd dropped out of law school as well. Erik explained that as a son of a college professor, it hadn't been the easiest decision. But he was sitting on a tram on his way to an exam in law school when, he recalls, "the penny dropped." He realized that if he passed the exam, he would most likely become a lawyer, and this realization frightened the hell out of him: *This shit is all I'm ever going to do for the rest of my life.* He couldn't shake the feeling that he wasn't destined for law, and after five minutes of sitting for the exam, he stood up, walked out, and got back on the tram. During the ride home, another realization struck him. "I saw all the small houses built for working-class people," he remembers. I thought to myself: *This is going to be my future now, because I'm uneducated—not that there is anything wrong with that, but I want something else.* This thought was never far from Erik's mind in his early days working at BinckBank. To this day, he says he can point to the exact house he was looking at when he had this epiphany, when he also made a vow to work harder than anyone—college educated or not—so that he didn't end up in one of those houses. "That thought remained in my head for fifteen years," Erik says. "I told myself every day: 'Just keep going, keep going, and keep going and give it all.'"

Like Chris, Erik wasn't cut out for traditional learning environments, but that didn't mean he wasn't extremely intelligent. Growing up, Erik was endlessly curious about how things worked, and he had little patience for those that didn't question things. Moreover, he disliked everything he considered to be *stupid*. What he meant by stupid, he explains, was rote memorization. "There is a difference between remembering something and *knowing* how things work together," Erik clarifies. Erik wasn't satisfied with memorizing. He could not only read the words of the books in his Greek and Latin courses but could understand them and make comparisons and analyze them as well. "Which is," he says with emphasis, "the most important thing."

Erik's ability to think on a deep level, analyze how things worked together, and be insatiably curious about how things worked made him a valuable asset to the BinckBank team. He was also willing to ask questions over and over again until he got the answers he needed. When something went wrong and everyone looked the other way or went home, he and Chris wanted to see how things worked, and both were willing to work through problems. Chris remembers when the system at BinckBank went down for fifteen hours one day; it was mayhem. "When I say everything was down, I mean *everything* was down. No systems, no internet, no phone, email, connections to stock exchanges, no nothing—from early morning till late at night." Clients came driving in from all over the country to see if the company still existed. Nervous BinckBank clients even called the Dutch Central Bank and the press. There was panic everywhere; hundreds of thousands of investors could not get to their portfolios and buy or sell their shares. When the systems finally worked again late at night, all hell broke loose. Claims from clients came rolling in—millions of euros' worth of claims. When Kalo asked who wanted to come and help out during the weekend to figure it all out, 99 percent of the people looked the other away, however, Chris recalls. "Erik and I were like, 'Hell yeah, we are there!'"

This underdog status fueled their motivation and drove them to figure out problems faster and work harder.

Bonded by their curiosity, tireless work ethic, and drop-out status, Bas, Erik, and Chris felt they had a lot to prove and worked harder than most of their college-graduate peers. In fact, this underdog status fueled their motivation and drove them to figure out problems faster and work harder, and they were the first ones in and the last ones out of the office for eight years straight. "One day," Chris recalls, "we had a problem with the systems, and Bas was at the dentist. We called the dentist and told them to put Bas on the phone. As soon as he heard what was going on, he left the dentist with only half his teeth done."

They each learned firsthand—by doing it themselves during these years—how to build a bank, by understanding and learning the banking business, customers, and processes inside and out. They also learned how to adapt, change, and grow. For example, while working at BinckBank, the retail business did not grow fast enough in the early years. It was then that Kalo asked Erik and Chris if they could build something that a professional investor could use to invest for several people at the same time. Chris said, "The only thing we needed to build was an extra layer, so that the professional investor could see all the accounts he managed (instead of just one account in retail), and they could do a trade on each different account." That was the first business process outsourcing (BPO) solution they built. Then they started enhancing it so clients could place an order for several accounts at the same time or have an overview of all the different trades. When it was time to sell the BPO services, Chris went with Kalo to present their features to asset managers. There were already two large banks in the Netherlands that offered the same services, but Kalo decided to offer this service at rock-bottom prices as well so that asset managers and banks would have the incentive to switch. They also wanted to make sure the product was ten times better so that when professional advisors logged in and used the system, Chris hoped "their jaws would fall to their knees and they would never want to use anything else ever again." In short order, the B2B operation became too large too quickly for Chris to manage in addition to the retail market on his own, so Kalo hired someone. Together with Erik, the new guy, Joost, ran the BPO team. Erik was charged with building and constantly upgrading the core banking engine software that BinckBank ran on to service these professional investors, with the ultimate goal to service a retail bank.

Meanwhile, Bas, who got his start in a help desk role at Getronics and had moved up internally to junior systems engineer, senior systems engineer, networks, security, managing teams, knew infrastructure inside and out. "What makes Bas so different from other CIOs is that many have never touched a server or installed one in their lives," Chris says. "Bas, on the other hand, has had his hands on all of it."

At BinckBank, Bas was responsible for the infrastructure of the servers. He built up BinckBank's IT division, where he was responsible for everything from networks to security and from Windows servers to Oracle databases and IBM RS/6000 machines. Then his responsibilities broadened to encompass team management of what was to become one of the biggest financial shared service centers in the Netherlands. So he went from helping people who could not log in to running a real-time trading system with millions of trades and connections to stock exchanges all over the world that had to be up for 24/7, 365 days a week with hundreds of thousands of retail clients logging in at a time. "That is what I call experience. Theory and practice are many times two totally different things," Chris says.

Erik and Bas worked closely with a developer/coder, Ilco van Bolhuis. Ilco, the son of a potato salesman (his father traveled to potato farms, filled his truck with crates of potatoes, and delivered them to buyers), grew up wanting nothing more than to get as far away from potato farms as possible. This desire led him to the world of electronics and computers. At first he studied and received certification to be an electrician. Then he studied computer science at the University of Amsterdam. "If I start something, I finish it," Ilco declares proudly. A man of few words, he likes to get to the point. He took to coding quickly and figured out he had a willingness to work through problems until completion. This was a skill that came in handy at BinckBank, where there was no end of problems to solve in the IT and BPO departments. When he started, he worked primarily on developing internal systems but then moved on to Erik's BPO team.

By the time Matthijs joined BinckBank, Chris had made quite a reputation for himself as one of the most knowledgeable yet instinctual employees at the bank. Matthijs, however, brought something different. Matthijs's curiosity to know more and not stop digging until he found what he was looking for was what propelled his initial research work at the bank. While working on his thesis, he interviewed over nine hundred BinckBank clients to learn about their trading behavior and whether or not they listened to or accepted advice from equity analysts. His work and research was covered widely and put Matthijs on the financial world's map. And the work he did at BinckBank proved he was a natural. He was great at talking to clients, managing portfolios, and fixing problems. Chris remembers, "During his internship, many times we asked him to pitch in and work with us. We told him to work on his thesis on his own time. With no work experience at all, he did all the things we asked of him, and he helped us during a time of rapid growth." It turned out Matthijs was more valuable than just providing access to his college professor's connections.

Over the course of the next several years, teams shifted, and sometimes the men worked independently of each other. But no matter where they were in BinckBank, they each gained a deep and broad level of knowledge of how banking and the core banking engine platform of banks work from the inside out.

LESSONS LEARNED IN IT

When BinckBank started in the beginning of the year 2000, the IT team had procured an on-premises version of a back-office application called Europort, an engine that processed the transactions that were coming through the website. There was also an add-on application called Europort Internet Brokerage (EIB), a form of middleware, which the team used to connect the website to the back end, where all the transactions were processed. Three years in, the supplier of this platform came up with an "improved" version called Europort+. Europort+ basically made the database structure more open for BinckBank to use (ideally to get more information out of it), and it promised to run processes better and faster. What this meant, though, was that BinckBank's IT team had to migrate from the Europort version to the Europort+ version. Seemed simple enough, but it was a massive undertaking. It also happened to be the first real application migration that everyone—Chris, Erik, Bas, Ilco, and Matthijs—worked together on as a team. However, at first, before being handed over to this team, the project had been run by a different IT director. But a couple of months before the deadline it became clear to everyone at BinckBank that *nothing* worked. The project was deemed a "disaster," like many others during these years of fast growth.

Kalo told Chris that he needed him to intervene and salvage the migration—in less than four months' time. So Chris drafted his plan and composed his team. Back then, Matthijs was only one year into BinckBank and, as he admits, "was still wet behind the ears." So Chris took on this project while there was, as Matthijs remembers hyperbolically, "blood streaming through the streets." Chris then structured the project and gave it a solid plan. "He was like Hannibal from the A-Team," Matthijs recalls fondly. Chris pulled everyone together, composing the team based on everyone's strengths and skills. Then he instructed them all to report back to him on a daily basis. "It was like a military operation. And at the end, even though the migration itself turned out to be a success, it was not a bump-free ride," Matthijs remembers. "It was painful, but not half as painful as it would have been if the old team was going to do it."

This migration was the first real application and database migration that Chris, Bas, Erik and Matthijs did in their lives, and they learned the hard way how to do it—through experience. And secondly, it was through this process that the two created what Matthijs calls the "blueprint of all the important ingredients in a project." The blueprint was and still is the following: *scope determination, governance, decision risk logs, priorities, phases, and meeting structures.* "Everything that we did back then we still do, only much more strictly than back then. Today we say, 'Don't change the scope—period!' and 'Give the team the mandate, and only decide the huge things on board level,' and more stuff like this. But it was at BinckBank that we did it for the first time," Matthijs explains. "As IT guys, we are very black and white during these projects, but that is a good thing."

Five years after that migration, Chris and Matthijs also learned from another major experience—another big technology upgrade within BinckBank. BinckBank's first website was built on Microsoft's ASP framework. But .net was introduced, and they wanted to adopt that technology. This was something none of them had ever done before. At first, for this migration from ASP to .net, they decided to outsource the work to a firm called ISDC in Romania, where labor was less expensive than in the Netherlands. But when the company came back to them several months later and presented their work, nothing worked. Chris says, "This was not entirely their fault, of course. We continued to work on the site during these months, so of course nothing worked when they delivered the code back to us. We did not have a good way of working with external developers, so we also learned this the hard way."

The team back at BinckBank had to start all over and threw everything the Romanian firm had done in the garbage. They rebuilt the entire website from scratch in .net. "We worked agile before agile was a method," Chris recalls. "We worked in small teams in two-week iterations with clear priorities and upgraded the pages on the go. It was an amazing project, and in a couple of months, we rebuilt the entire BinckBank website and BPO site in a new language and made it ten times better along the way."

It was clear that they had a deep understanding of team composition, governance, and clear decision-making. "Our thinking was this: 'Leave us alone; we'll manage it and will deliver it,'" Chris says. "We went in a room and decided everything ourselves, and we delivered the best financial website the Netherlands had ever seen with a team of less than ten people. We were working agile before it was a term."

Because it was Matthijs's and Chris's first experience handling nearshoring or outsourcing—managing contracted teams outside of their office—their learning curve was steep. It was during this project that they learned what not to do. Chris and Matthijs ended up managing the project. There was a whole host of obstacles to overcome in order to make sure the two were compatible and secure. This experience, however, was invaluable.

A PROBLEM LEADS TO AN IDEA

Binck had other outsourcing issues as well. At one point, Binck was having a number of technical problems with its core banking software, Europort. Binck thought the vendor should pay for the fixes, and the vendor thought Binck should pay for them. Binck wasn't a top priority for the vendor. In their mind, the vendor had other clients to attend to and more pressing matters to address. Kalo wasn't having it and decided to buy that company so Binck would most definitely get top priority. However, after Binck bought the vendor and the team started asking them to make changes to the software, the vendor still pushed back and said they had a lot of other clients with other contracts and they couldn't get started. Needless to say, it was, in Chris's words, "a very strange time" because they owned that software company and the company still was not making the changes they thought they should be making. Chris suggested all these changes to make it better. It was really bad software. But they did not want to change anything. The only answer was always to put more hardware in the racks, but this did not make any sense at all. It was then that Chris started making plans to evaluate processes and determine which process had priority over the others. It was also then that Chris said, "To hell with the vendor!" and that the changes should be made internally. Chris remembers telling Kalo, "We should start reprogramming certain processes and elements of the software that are no good, especially for real-time trading, real-time transactions, and speed. We have to make a loosely coupled system where processes can work independently from each other so we can also use server power exclusively for a certain process." It was there that the seeds of a bigger idea were planted. "My mind just started working on a new system—even back then—because the software vendor that Binck bought and that later became Ohpen's biggest competitor in the home market didn't want to make the changes that were needed for the online, real-time, and mobile world we were going to live in."

It was then they began to see that Ohpen team's proverbial IT-software-upgrade/migration toolbox was starting to get filled. Additionally, they learned about the ins and outs of vendor relationships. When all was said and done, Chris and his team at BinckBank did over twenty upgrades/migrations of their own systems and eighty implementations of asset managers as well as several insourcings of banks.

Over the course of eight years, they managed anywhere from fifteen to twenty-five projects simultaneously, and they did so 365 days (and nights) of the year. After years of working like this, Chris reflects, "We had gained more experiences in those eight years than most people do in a lifetime. We grew from a six-member to a six-hundred-member staff and from having no clients to having half a million clients, and we insource multiple other banks on our platform. We went from zero transactions to twenty million transactions. We went from complete anonymity to being listed on the stock exchange with a unicorn valuation (valued over a billion dollars), and we did it all with a team whose members were almost all under the age of thirty. We went from the Netherlands to Belgium, to France, and then to Italy. We went from running our own bank to insourcing and running other banks at the same time, and from upgrading systems, networks, servers, and software, all while running a fast-growing operation. We went from broker to full-fledged bank with all the regulatory licenses. Though many of these things are common today, we were the first to implement real-time ordering, real-time quotes, real-time streaming of quotes, and real-time news online. All of this was done for the *first* time by us."

It was at BinckBank, the team members realized, that they gained more than just experience. They gained confidence in their abilities as well. As Chris says, "Sometimes all hell broke loose, but we all loved it. Kalo and Thierry Schaap were the founders of the company, our leaders and mentors, and without them it would all never have happened, but we were the team on the ground running the day-to-day operations. And they let us do that. That was pretty cool of them, because we did not really have the necessary credentials to do it, but we did it. We all managed to dive into the deep, with lots of currents everywhere pushing against us, and we all came back up without any scratches. Well, maybe one or two scratches."

LESSONS IN BANKING, RISK MANAGEMENT, REGULATION, AND COMPLIANCE

In addition to the technical and back end processes of running a bank, the team also learned several nuances within the banking industry itself, namely security, compliance, regulation, risk management, and even the technicalities of how to set up and license a new bank. When starting the bank, Kalo and his business partner, Thierry, needed help from the team to prove that the bank was compliant in everything from administration of the organization to operational procedures, including IT general controls, security, and application management. Matthijs and Chris had a front-row seat to watch (and even participate in) this Herculean effort, which included some two-thousand-plus tasks. Chris adds, "The funny thing is, every person who has not worked in a regulated environment has no clue what that means. We did the work of our own clients for ten years before we started our own software company. We all worked at a bank for ten years. We sat at the other side of the table for ten years. So we all made mistakes, of course, but at the end of the day we know what we are talking about, because we have been working in the same regulated environment for so long."

In 2003, when the DNB finally granted BinckBank the license, Kalo called Matthijs to tell him the good news. Matthijs was in London visiting a friend and remembers the conversation to this day as much as the feelings of both elation and relief that accompanied it. The licensure was not only a win for BinckBank; it was also a historic win as well. Other than foreign banks with branch offices in the Netherlands that had been awarded licenses, BinckBank was the first bank to get a *brand-new* license in nearly two decades—and by a young team to boot. When the team went to the DNB to pick up the license, the people manning the door of the bank turned the young men away. In fact, they told them to go farther down the road a couple of blocks—to the City of Amsterdam municipal offices, where they could pick up their *parking licenses*. No one could believe that a group of guys in their twenties was at the DNB to actually pick up a *banking license*.

LESSONS IN BECOMING A UNICORN

When BinckBank was founded, Amsterdam Option Traders (AOT) held a 51 percent stake in the company; the rest was held by the founders, Kalo and Thierry, and some employees, among them Chris and Matthijs. In May 2004, AOT said it wanted to buy the remaining parts of the company. They paid the founders and the personnel not in cash but in shares of AOT at a price of €1.18 a share. AOT had capital but no business at the time, because their technology was not up to date and they weren't able to keep up with market demand. AOT shareholders recognized this and went to BinckBank and asked them if they could take over the whole company. With their remaining cash, AOT bought BinckBank, and BinckBank became the parent company. In short, it was a reverse takeover. Since AOT was already a listed company merging with BinckBank, BinckBank was now a listed company. Chris, Matthijs, and other BinckBank employees were given stock in the company at the inception of the company, so when it went public, the shares had value. But one of the principles of the reverse takeover was that there was a lockup on the stocks to ensure that shareholders didn't sell them outright. They could only sell 25 percent each year. Shortly after this, Matthijs and Chris took a vacation to Costa Rica. They stayed in small budget hotels, rented a Suzuki Swift to get around, and even shared a room to save money. One morning, while they were staying in San José, Matthijs got up early and found an internet café to check the news and review the stocks. When he got back to the hotel, he told Chris the news: BinckBank's stock had more than doubled in a couple of days—it was now €3.30. Matthijs, who Chris says was usually pretty contained and not easily enthused, was positively jubilant. He couldn't believe the rapid growth. The two young guys living on a budget were now "rich." Though Chris recalls, "On paper we were worth a quarter of a million dollars, but because of the lockup, we couldn't sell." Chris readily admits that though they were initially upset by the lockup because it meant they couldn't sell their shares (joking together: "Screw the fucking lockup!"), they now realize it was the best thing that ever happened to them. Because of it, the two self-proclaimed "young and naive" guys couldn't sell their shares. Over the next three years, BinckBank grew exponentially—beyond their wildest imaginations. "But it was a strange switch—from a very open company where everyone knew and heard everything, to being listed on the stock exchange where you can't just say everything to each other anymore. That's a big change in culture."

LESSONS IN EXPANSION

Then, in 2006, Kalo asked Matthijs to look into expanding BinckBank into the foreign market. Matthijs did his research and went back to Kalo with his results: the place to start was France. Even though BinckBank had a banking license in the Netherlands, they still needed to be authorized by the Bank of France (Banque de

France). Since the road to licensure was well paved by then, Matthijs simply needed to make sure that everything was complete to get the branch-office recognition by the Bank of France. Matthijs, at just twenty-nine years old, under the supervision of Thierry wrote a lengthy document (nearly seventy pages) that outlined everything that BinckBank had in the Netherlands and how he would apply these processes and procedures in France to comply with all the rules of France's prudential regulator. Though in charge of all the research, Matthijs wasn't on the short list to be selected to lead the new office. But Chris convinced Kalo to give Matthijs a shot at setting up the new branch in France because he thought he could do it and because he deserved it.

Matthijs dived in headfirst and covered every detail to set up the bank. To this day, he says, "If you wake me up at three in the morning and you tell me you want to start a bank in the Netherlands or even in France, I can tell you immediately, from experience and knowledge, every single point that needs to be in there and what evidence you'll need to demonstrate to the authorities that you are fit to start a bank." Not only did he memorize the list of compliance regulations; he could explain in detail how the banks comply with those regulations.

The knowledge and experience Matthijs gained working at BinckBank was invaluable. When he talks about onboarding banks to the Ohpen platform, he can tell people exactly what they can and can't do to remain compliant. According to Matthijs, working at BinckBank was like attending "banking college." He could argue that all of his colleagues at Ohpen now are in what he deems the "Ivy League of core banking engine colleges." Ohpen employees are learning at even a higher level and are surrounded by other high achievers, all emboldened with a great sense of responsibility and accountability not to mention an eagerness to learn. They were the first team ever to put a bank in the cloud, so the learning curve was enormous. "We give them chances to run projects autonomously. We received the same opportunity at BinckBank. We were all green as grass, wet behind the ears, and yet still empowered to run important projects. We all fell but got back up again. *We failed forward,* as they say. All of it was new back then, though. Our Ohpen team now benefits from all that we learned firsthand back then."

Agreeing with Chris's earlier assessment that there is truly no better teacher than experience, Matthijs adds, "You can't learn all this from books alone, but doing it yourself really is more important. It all just sticks." As a parent, Matthijs stays up to date on the latest in child psychology and educational pedagogy. He read once that parents should let children learn by experience, unless of course they are 1) a nuisance to their environment, or 2) if whatever it is they are doing is life threatening. "What we did in those ten years at BinckBank was neither a nuisance nor life threatening. We weren't putting the interests of BinckBank's clients at stake. *We learned by doing.* We didn't fail on an epic, life-threatening level because we were driven and ambitious. We had the necessary guidance every now and then from Kalo. But it was a pressure-cooker education to get to adult life as quickly as possible."

LESSONS IN *WHAT NOT TO DO* WHEN SETTING UP MULTITENANT PLATFORMS

Meanwhile, as Matthijs was researching and setting up a bank in France, in that same year, 2006, Friesland Bank, a midsized bank from one of the Dutch northern provinces, decided to outsource its complete securities administration to BinckBank. This posed an interesting new conundrum—and learning experience—for the BinckBank team. BinckBank now not only had to administer transactions for their *own* retail clients and asset managers; they had to service Friesland Bank's customers as well. As Chris says, "It was one thing to manage our own retail customers and asset managers, but to insource another bank was a totally different ball game!" To do this, BinckBank had to transform its core banking engine from a single-tenant platform (just BinckBank and asset managers) to a multitenant platform (one that served both BinckBank and Friesland Bank)—in other words, a platform where two or more banks shared one basic platform and one administration. The obvious question this scenario proposes is the following: How does a platform that up until that moment in time served one bank transition to a multitenant platform? More importantly, how does a platform separate each bank's customers' assets—and keep them safely secured from each other? The BinckBank BPO team had to do just that. Erik, who was on the team, recalls that building the platform and figuring out a way to keep the assets separate was the easy part. They figured out a simple solution to give each client an identifier—so that all the information for each bank's clients had its own code. Seemed easy enough. However, this approach introduced some inherent risks. It relied heavily on humans (in branch offices) inputting the right identifiers. So they needed to run multiple tests to ensure that Bank A could not see or access any of Bank B within the platform. One of Erik's and his team's jobs was to mitigate these risks, and they took it very seriously. At the time, they did the best they could. But they were also learning as they went along. "We can say now, looking back, that the system was not a hundred percent secure; it was ninety-nine percent secure—but that's not good enough for us at Ohpen now. Because there is always the risk that someone can see an item that has been labeled incorrectly," Erik reflects.

In fact, at one point another bank did receive incorrect statements from BinckBank. Chris and Kalo had to apologize to the client and assure them it would never happen again—only it did. When the client looked at their statement and said, "Hmm, this isn't a client of ours …" Chris and Kalo just looked at each other and didn't have to say anything. They both knew what the other was thinking—it was an imperfect system. As Erik explains, "It's like putting all of your clients' money in bags with their names on it. Then you combine all those bags of money and put them into the bank's safe. Same building (platforms), same safe (the banks). The problem, of course, is that the banks in the safe have access to not only their own 'bags' but also the bags of other banks. They can, even if it is a mistake, grab the bag of another bank client's money if it's mislabeled." The solution to this: Erik, Chris, Bas, and Ilco figured out that what they *really* needed to do was build a platform with compartments—a sort of siloed multitenant platform—so that the clients weren't in the same database. The risk was too high to have all the clients in the same database. "If it would have gone wrong, we would have been out of business," Chris concedes. "We did not have the scale yet to make mistakes like this and survive, so that is why we had to think of something else: a multitenant system but with separate databases for clients." Or to continue with the safe analogy, like creating safe-deposit boxes within a bank's safe. Each client has their own key to their own safe-deposit box/customer assets, and no one else can get at them or mistake them. Each can come through the same doors of the bank building and safe (one singular platform), but then they *each* need their own key to enter their own safe-deposit box (access their data).

While figuring out and troubleshooting the multitenant platform seemed easy enough, what was difficult was migrating the data. Erik shakes his head remembering the stress of the initial migration. "We started after working hours on a Friday afternoon so we could run it during the weekend when the bank was closed. The script was running and running and running. Then Friday turned into Saturday. And Saturday turned into Sunday. And by Sunday around lunchtime, we got nervous." All banks open on Monday morning, and Erik and his team weren't sure if the script would be done running, or if the script would fail and they would have to start all over again. They were far past the point of no return—when they couldn't just stop and go back to business as usual. Everyone was panicking. It was then that everyone learned something invaluable. Migration had to be done quickly—with as little downtime as possible. Though they knew at most banks this would be nearly impossible. The performance of most legacy systems was poor and the systems themselves so old and outdated that no matter how great the new software, it would be nearly impossible to migrate efficiently. As long as these systems were in place, it would be a nightmare when it came time to migrate data. They realized there had to be a way to run upgrades and migrations fast, not just so downtime for banks would be shorter but because, Erik adds, "You need some kind of safety net (extra time) if shit hits the fan."

It was at BinckBank that Chris had the idea to make what he saw as a "freeway." He noticed that most clients didn't care if things took a couple of seconds longer. He adds, "It does not matter if a lot of processes take ten minutes longer or even an hour longer. At BinckBank I started thinking about creating a bank system where we could prioritize processes and allocate computing power to certain processes. So we could actually guarantee that a process was done within a certain time window. Now, fifteen years later at Ohpen, we are still the only core banking provider in the world that guarantees how fast processes are executed."

Erik, Bas, Matthijs, Ilco, and others started to question how platforms should be built from the start. Each one of them was searching in their own way not only to make the processes better but also to make the platforms they were using—and the servers/systems they were sitting on—not just better but faster, more agile, more compliant, more reliable, and above all safer. "I think if our clients could actually look under the hood of the car, they would be amazed at what we have made," Chris says.

HANDLING THE COMPETITION

At the time, Binck's biggest competitor was Alex.nl, where Chris had worked for two years before joining Binck. Alex.nl was bigger than Binck and already had a sizeable market share. "Alex was always just two steps ahead of us," Chris recalls. "We were looking at them and they were looking at us, and they didn't like us because we were gaining market share really fast, because we had better systems, better customer service, and lower prices. We had more drive than they did, and we were going to crush them. That was our mentality."

Not wanting to miss a thing and wanting to keep up with the competition, Binck planted a "spy" at Alex.nl. Binck sent a half brother of one of the key people at Binck (they had different last names) to do an internship at Alex.nl. One day the intern came back to Binck and shared that Alex.nl had been developing a top secret program for their most active clients, who were responsible for a huge part of Alex's revenues. The intern reported that Alex was going to launch an upgraded trading feature and creating a system that would meet just their top clientele's needs. When the team at Binck heard the specifics of what Alex.nl was doing, they were flabbergasted. *How could we have been so stupid? Why didn't we anticipate this? Why didn't we think of this?*

When Kalo heard this, he looked at his team and said, "Okay, let's build something like it." Chris said they could, but it would take a month or two of around-the-clock work. He made Kalo an offer: "If you leave us alone, let us do what we do, give us a little bit of a budget so we can have some sushi instead of pizza, we'll work from seven in the morning until midnight, seven days a week, until we have something that can compete." Chris also tempered Kalo's expectations. "We can try to build the system like Alex.nl's, but to be totally honest, when we're done we'll have something quite good, but it won't be as good as their system because they had fifteen months to work on theirs and we had two."

Kalo didn't care about its being better than Alex.nl's; he wanted to get to market first, and he wanted to make sure Binck's system went live before Alex's. Then there was the issue of price. Since Binck was already a listed company, lowering the prices for this group would have made a huge impact on Binck's revenue. However, the Binck team couldn't get the data they needed to start making calculations and determining what the impact would be for lowering the price for this select group of clients—again, because of the crappy back end system. Kalo's recommendation was to hold off and see what Alex. nl's price would be and then offer a slightly competitive price. Ultimately, Kalo's plan was to go live the exact same time as Alex and to price it on the fly as soon as Binck found out their numbers.

One morning, the intern reported Alex would be going live the following Saturday. The Binck team was ready. However, Kalo was already scheduled to be out of the office for a long weekend in the South of France with friends.

That Saturday morning the new feature was to go live, Chris had to call Kalo some fifteen times to wake him up. When Kalo finally answered, Chris said, "We have to go live now. The press release has to go out now. So what do you want to do with these prices?"

Kalo asked what Alex's price was for options.

Chris replied, "Three euros."

"Make ours two point seven five," he advised. Then asked Chris, "What about the futures?"

"Three euros," Chris said again.

"Make it two point seven five," Kalo said. "Now, what is the price for the stocks?"

"It's ten euros plus point one percent," Chris replied.

"Make it nine euros plus point zero nine percent," Kalo said.

"Are you a hundred percent sure? Because we have no clue what these prices will do to our revenue," Chris asked again.

"Do it. Go live. I'll call back later," Kalo said, quickly hanging up and going back to sleep.

That following Monday, all the newspapers broke the story: *Binck and Alex both launch new trader application.* When the boss at Alex saw this, Chris says, "He went completely mental." Chris knows this because years later he confided to the team that he'd heard that the boss was throwing stuff all over the office that day. For the life of him, he couldn't figure out how Binck had found out they were building this application, let alone when they were going to go live with the press release. Chris recalls triumphantly, "I think in his whole career, this was his biggest annoyance. And, yeah, we laughed our asses off. It was brilliant, and it showcased the entrepreneurial spirit that was at BinckBank those days. After Kalo left, that culture was gone … and the company with it."

BINCK'S FINAL DOMINATION OVER ALEX.NL—THE MARKET

Then, in February of 2007, BinckBank decided to buy Alex.nl. This meant that two of the biggest broker platforms would merge—dominating the market.

In order to buy Alex.nl, BinckBank needed to pay close to €400 million. BinckBank didn't have that money, so they needed to raise capital. They thought of a solution: approaching existing shareholders through what is called a rights issue, which is a group of rights offered to existing shareholders to purchase additional stock shares (called subscription warrants) in proportion to their existing holdings. In a rights offering, the subscription price at which each share may be purchased is then discounted relative to the current market price. These rights are often transferable to sell them in the open market. So BinckBank said to investors, more or less, "For every share that you have, you can buy one more share at a discount of three euros." This offering was wildly successful. They not only raised the capital, but when all the dust settled and BinckBank merged with Alex.nl, the price of everyone's shares soared—making Chris, who owned a considerable number of shares, a very wealthy young man. At just thirty years old, Chris had a net worth of nearly €1.2 million. However, prior to the rights issue, Chris decided to take out a loan to buy extra shares—using existing shares as collateral. At the time, he thought he was essentially doubling his money. The only thing that would stop him from growing more capital was if all hell broke loose and the stock market suddenly collapsed like it never had before.

CHRIS LEAVES BINCKBANK—BUT NOT
THE EXPERIENCES—BEHIND

Not long after the Alex.nl takeover, in early 2008 Kalo decided to leave BinckBank. At this time, he had a chat with Chris about Chris's future at BinckBank. He took Chris aside and said, "There is only one person here qualified and knowledgeable enough to run this bank after I leave, and that's *you*. And there is only one person they can't give this job to, and that's *you* too." Chris knew immediately why. Many people making the decisions in the company were against it. Chris was like a mini Kalo, and Kalo did what Kalo thought was best for the customers and did not listen to anyone. The board of directors did not want another Kalo. They wanted someone who would do what the board wanted. So before Chris was even considered, he was off the list. Chris decided he needed to make a change, especially because his plan of reprogramming the Binck systems got a no from the board. He said on his way out, "Mark my words: we are worth one point two billion euros, and in ten years' time, that billion will be gone from the valuation." Chris admits cheekily that he was "wrong" about his future-casting—*sort of.* Instead of losing €1 billion, they lost €800 million, and BinckBank was sold for only €400 million ten years later. Chris left BinckBank on a high note. He took his €1.2 million and decided to take the next few months off to travel the world with his girlfriend and figure out what he wanted to do with the rest of his life.

In the meantime, Matthijs was running BinckBank's French branch office. While Chris was turning inward and focusing more on yoga and meditation to manage his anger in reaction to those who'd treated him with disrespect or who'd triggered his previous trauma he'd experienced in his youth, Matthijs was learning about his own leadership style on the job. He was not only building a bank but testing his own mettle as a leader. "I literally started from scratch. I had to ask, 'What kind of office do *we* want? Do *I* want? Where should it be? And what should the vibe of the place be?' And I needed to start working on a team, because obviously I couldn't do it all by myself. So I asked myself, 'What type of people do I want to work with? Where are my future customers? Where are my suppliers?'"

Eventually, Matthijs selected a suburb of Paris and found a suitable office. He also hired his first employees—legal and commercial experts. He cast his net wide and tried to build as diverse and multicultural a team as possible—which was fairly easy to do in Paris. He wanted divergent and innovative thinkers with a variety of world experience. He also wanted to consciously create a team with both men and women. But regardless of race, religion, or gender, at the very core of his hiring practices was a desire for competence. *Could people deliver and do what they said they would? Could they give it all, exceed expectations, and do more with less*—all the things he and Chris valued so much when they had worked together? "It was for me, this incredible experience—my first time—to build my own team and determine what kind of leader I wanted to be."

Though at first Matthijs was not even short listed to be one of the people to go to France, Chris had persuaded Kalo that Matthijs was the man. Chris knew Matthijs wouldn't go out of his way to tout his own achievements or skills. So Chris stepped in and did it for him, and Kalo agreed to give Matthijs a shot at it. In the end, Matthijs exceeded everyone's expectations. For Matthijs, the qualities that served him well in school, then as a BinckBank intern, and then a bona fide employee—grit, determination, curiosity, and a willingness to keep digging until he found what he was looking for—served him well as a leader in those early days in France. These skills combined with his unique ability to think through complex issues with a calm and even temperament helped too. He eventually got the bank up and running and in no time became the number three online broker in France. Overall, it was considered a monumental achievement and led the way for future branch openings as well. In fact, the French branch opening was the highlight of BinckBank's 2008 annual report, which featured photos of Matthijs's accomplished and diverse team posing all over the streets of Paris. BinckBank had gone international—and Matthijs had taken them there. For Matthijs it was not just a professional highlight but a personal one as well. He loved living in Paris and fully embraced the joie de vivre.

Meanwhile, Chris's extended family back in France, who had seen the commercials for BinckBank on French television, finally realized (or were ready to admit) that Chris's growing up in the Netherlands hadn't been such a bad move after all.

LIGHTNING STRIKES

In the meantime, while Chris was exploring the world, he couldn't help but keep thinking about all the tech problems he had been running into while at BinckBank. He knew there was a better way to build software. He did not want to be "that guy" who criticizes things but does nothing about them. He had criticized the core banking software at BinckBank, and he wanted to prove he could do better. Chris recalls, "I knew the idea that formed BinckBank to start coding a new retail bank from scratch was a supercool idea. How many times in life do you get the opportunity to actually start from a blank sheet of paper?"

While traveling, Chris went to Africa, Asia, Australia, and Fiji. All the while he was reading newspapers, talking to locals, and just seeing what else was out there in the world. After being in the office for eight years straight, he wanted to open his eyes and see and experience different cultures, different ways of life, and different ways of looking at things in general. He was actively looking for what he calls a "brain shift." He recognized that after being in an office for eight years straight he may have acquired what he called "tunnel vision." Chris adds, "I traveled to open my eyes and make my brain work differently. On all these travels, I was seeing things and meeting people, and they all had different ways of looking at the same thing." Chris was also able to read local newspapers that he couldn't get in the Netherlands. (He once read that Warren Buffett read for four hours every day and that he reads several newspapers.) By the time Chris had arrived in Los Angeles and San Francisco, he was well versed and up to date on what was going on in the world.

When he arrived in California and started reading the local newspapers, watching the news, learning about the cloud, things started to click. It was as if he was again unplugging from the matrix. He could see it all and how he could build a new core banking engine from scratch that was reliable, adaptable, and compliant and that would be built with unlimited scalability in mind and could be easily integrated with large IT landscapes—using a cloud-based server rather than a legacy-server architecture and that would be infinitely scalable and fast: SaaS for banks.

Chris's rationale for using the cloud was multifaceted. "In the old days, you had to buy a server, wait two months for it to arrive, unpack it, configure it, put it in racks, then test it, connect it to the network, and so on. And then after all that you had to upgrade it and maintain it, run patches, and so on. I just saw that this cloud thing did a lot of work that we normally had to do and that it would cut our IT staff at least in half," Chris says. The second thing Chris liked about the cloud was that it was automatically superfast, could operate without mistakes, and could scale up and down automatically. This would cut another 15 percent of staff. The third thing was that it all worked on demand, which, he says, saved him another 5 percent in staff. And the fourth thing was that he would only have to pay the cloud provider Amazon Web Services when he used them. "This is a mega thing," Chris says. "In a way, we're kind of renting them instead of buying them. This could change our balance sheet as well. I knew the load on the servers of the bank, so if we could only use and pay for the times when we used them and take them down (automatically) when it was night or when we knew no one was using the system, that would further cut our costs enormously. And the idea I had a long time ago to allocate a certain process to computing power was now a lot easier to do. All of a sudden we could write the software specifically for the cloud, and that would make us very scalable, fast, and effective." He could onboard not one, not two, not three, but an infinite number of banks. He could take what he and his friends learned at BinckBank and build a better, faster migrating process as well. He could start coding a bank from scratch using cloud technology that would work all the time. IT. JUST. WORKS. He knew that would be its unique selling point. "If we would have had this kind of technology at BinckBank, we could have done it with three hundred instead of six hundred people. Imagine what this would have done to our cost income ratio. And our valuation!" He knew just the team to do it with, and he knew that if he could pull this off, it could become massive—a massive shift in how retail banks actually work.

> **He knew that if he could pull this off, it could become massive—a massive shift in how retail banks actually work.**

All he needed to do was get some funding and he would be ready to go. But Chris was in no particular rush. He had plenty of money to live on for years if he wanted to—he had over €1 million in the bank after all, and he was enjoying his life and all the time he was finally getting to spend with his girlfriend, Myrthe, after years of working incessantly.

The idea could take its time …

THE WORLD TURNS UPSIDE DOWN

But on September 15, 2008, which also happened to be Chris's birthday, the financial world came crashing down. Lehman Brothers, the global financial services firm, filed for Chapter 11 bankruptcy protection after massive and drastic losses due in large part to the subprime mortgage crisis. Within days, the entire financial services world was flipped on its axis, and global markets immediately crashed as well. Among those who lost their life's fortune was Chris—he lost nearly 85 percent of his net worth that year.

No longer buoyed by the cushion of fortune and time, Chris felt the urgency to act. As soon as he returned to the Netherlands, he began to pull the team together. After Matthijs turned him down in order to keep working at BinckBank in France, Chris, undeterred, began talking with Bas and Erik on a regular basis. Each of them, however, was still working at his job. Chris knew that if this idea was going to work, he needed an office and a team that could work full time. He wanted his friends to be able to provide for their families while they worked for him. If this idea was ever going to take off, he needed someone who believed in his idea—and in his team.

He needed an investor.

CHAPTER 3

GIVING AWAY A PIECE OF THE PIE

IF THERE WAS ANYONE IN THE WORLD who was a match for Chris's mindset and who could even understand what Chris was talking about—a core banking engine on the cloud—it was Michel Vrolijk, the managing director of Amerborgh. Amerborgh is a family office / private equity firm that specializes in investing in companies related to asset management, cross media, hotels, and properties as well as art and cultural events. The name Vrolijk literally translates in English to "cheerful" or "happy," and if there is one outstanding feature of Michel "Happy" Vrolijk, it is his cheerful if not positively vibrant personality. With a shock of white hair and a broad, gleaming white smile that spreads easily across his clean-shaven, handsome face, Michel seems to emanate an otherworldly glow of positivity everywhere he goes.

Michel, however, would be the first to say, "Don't let the smile fool you." He's indeed energetic, charming, happy, and lighthearted. That is all true. But he is just as rational, sharp, and fast thinking as the next venture capitalist. In fact, it's more apt to say that he has the rationale and measured behavior of a wealth asset manager, which is where he got his start. Before becoming a venture capitalist, he came from inside the banking world and knew it intimately—the highs and the lows and, yes, even the pains of legacy systems. He had worked in banks since he got his start at Merrill Lynch (now Merrill) back in the eighties. And during his work on the board at ING, he saw his fair share of legacy-system issues.

In 2006, when Alex Mulder started Amerborgh with just three people and asked Michel to run it, they, like Chris and his team, knew how to do more with less. Alex Mulder, now in his seventies, is one of the wealthiest people in the Netherlands. He started the first temp-work company in the Netherlands, kept the company his entire life, and turned it into one of the biggest companies in Europe.

He always invests in people more than in a business plan, and when Chris came in to the Amerborgh offices with all his tattoos blazing, Alex liked him immediately. His reason? *Not the typical banker—I like this guy.*

Alex and Michel also had a knack for seeing the genius in things that others could not. In fact, in 2006, just after opening their doors as a firm, a man walked into their office and said he had an idea to make the famous book *Soldaat van Oranje* (*Soldier of Orange*) written by World War II Dutch resistance fighter Erik Hazelhoff Roelfzema into a musical. The idea was ambitious. But Michel remembers listening to the man speak passionately about how he planned on telling the famous story of resistance fighters in the Netherlands during the German occupation of World War II. At its core, the story is about how one friend wanted to fight the Germans and the other wanted to finish his studies and do nothing. As Michel listened to the man's idea, he couldn't help but feel energized and saw something timeless and profound in the project immediately. Michel recalls, "I realized the dynamic in these young friends was a story of *today*. It's a story that may have happened in the forties, but it's a very important story to retell in a very new way for new audiences."

Many in the private equity market thought that Alex and Michel were crazy to fund such a project. But Michel and Alex were resolute. He and Alex began to pull in a variety of experts in lighting, setting, and even acting—spending nearly €10 million to get it up and running. In the end, what they created was something wholly unique and incredible. Everybody involved gave it their all and molded it into something that had never happened before in the Netherlands. It became the most successful musical in the Netherlands' history—it has sold out every show since its debut and has been running for ten years. Not a bad investment for his first time out of the gate as a venture capitalist.

A man of both instinct and reasoning, Michel wasn't just a visionary when it came to successful musicals. As someone who was working at Merrill Lynch on Black Monday, October 19, 1987, Michel had witnessed the mercurial nature of the stock market. He had a front-row seat for the collapse of not just the stock market but several companies and peers' livelihoods as well. It was the sort of trauma that stayed with him. He never forgot it. Some would say it stayed with him on a deep subconscious level. So when he was in the United States in the summer of 2007, traveling with his family in New York and California, Michel couldn't help but feel something was "off" or deeply wrong. "It was so strange," he recalls, sitting in his Amsterdam sun-drenched office and staring off out the window, "because even the local news was talking about financial issues. I'd never seen that in the States. It's always about the new kangaroo at the zoo or something like that. But now the local news had a totally different vibe around it." Something inside him clicked. Perhaps it was subconscious and he was remembering the precipitating events leading up to Black Monday. Or perhaps it was just a rational insight based on news reports that the mortgage markets were unstable and that banks were foreclosing on people's homes. Nevertheless, Michel was certain of one thing: *All was not well.* Almost as if it was a premonition (or like when Chris felt he was unplugged from the matrix and could see everything clearly), Michel felt he knew exactly what he needed to do. Prior to leaving for the States, he and Alex had been discussing investing even more of their capital in the stock market. However, after watching the American news, it became clear to Michel they should not only stop these plans but pull out *all* of their capital that was currently in stocks and put it all in low-yielding government savings bonds. He was certain that "whatever was happening in the States was going to be moving across the ocean soon." As soon as Michel returned to Amsterdam, he told Alex of his plan. He remembers saying to Alex, "Listen, I've been there, and I think something is terribly wrong there. Everybody's telling us we need to step into the market today, but I would now pull everything, take it out of the banks, and put it in German and Dutch government bonds. Everything we have." He then explained further, "I started in this business in 1985, and I have *never* felt this way."

Alex asked Michel one question: "How much is this going to cost us?" Michel's answer was honest: "I don't know how much it is going to cost. But *if* what happens is what I think is going to happen, the damage would be devastating." When Michel told others in the industry what he had done, again they thought he and Alex had lost their minds. They warned them, "You cannot make decisions this way, because you have to do what's rational and use models." Michel ignored their advice.

Shortly after pulling all their money out, Michel watched as the news broke: Lehman Brothers fell. Then the markets, then financial institutions, and then, in short order, venture capitalists were decimated. Everyone around them was losing their shirts. "The whole thing rolled over us. It was total and complete." Michel pauses and hesitates to find the word before concluding, "I don't know what *it* was, but it happened that way." Amerborgh not only managed to escape the tsunami that wiped out so many others; they ended up thriving. "As a matter of fact, these government bonds moved up dramatically, so it ended up being a very profitable decision."

All of this is to say that when Chris arrived at Amerborgh, it was one of the few firms left with actual capital to spare in early 2009. In fact, Chris couldn't have picked a worse time to start a company in the fintech industry. Both were floundering after the crash. Chris was undeterred. When Erik called Chris to report further dips in the market, Erik recalls Chris saying to him, "I can't think about that right now." Chris's vision was resolute. Besides, he rarely cared about what other people were doing and did his best to avoid negative or catastrophic stories. He knew what *he* had to do. He was determined. So when a mutual acquaintance recommended Michel/Amerborgh as a possible investor, Chris made an appointment, packed up his business plan that he had pulled together over the Christmas holiday, got on his scooter, and headed over to meet him.

Prior to meeting Chris, Michel had no idea who Chris was or anything about his background. In fact, Michel had no idea Chris had been personally affected by the market crash. Likewise, Chris had no idea Michel had managed to benefit from it. It was, in a way, a kind of modern "Miracle of Amsterdam" on the day the two met. Both men had literally escaped or survived the fires of life. Michel's company remained unmarred while the financial markets were burning, and Chris himself had survived the caldron that was his early life. The two made a perfect pair. Rather than refer to their meeting as a miracle, though, Michel likes to refer to their partnership more as a *Perfect Storm*. "All these events were happening at the same time—and to all of our benefits." It was, in his mind, the perfect time, the perfect place, the perfect idea, the perfect team, and the perfect leader.

> **It was, in his mind, the perfect time, the perfect place, the perfect idea, the perfect team, and the perfect leader.**

As someone who can see what others can't, Michel took a liking to Chris immediately. When Chris arrived at the Amerborgh office in the posh southern district of Amsterdam for the first time, Michel couldn't help but laugh a bit when he saw Chris show up on his doorstep. He was very young and had long, wavy dark hair and some visible tattoos. But, Michel recalls, he had also put on a button-down shirt and jacket for the meeting. Everything about Chris was completely unexpected, and Michel liked that about him almost immediately.

But still Michel was cautious. He was a rational investor, after all.

Chris came with the same trepidation. "All investors are smiling and charming at first," Chris says.

"We get a lot of people who come here with ideas, with plans, and sometimes they are very good, intelligent ideas, crazy ideas but …" Michel pauses. "Not all of them can answer my first question." Chris, Michel says, could answer his question immediately. "If I'm going to finance this idea today, what's *the first thing* you will do to execute this idea tomorrow?"

Most guys remain silent when Michel asks this. They don't know what they are going to do. "Some even say the first thing they're going to do is hire a consultant," Michel says with a hearty laugh. In this case, Michel recalls, Chris didn't skip a beat. He knew not only the first step—hire the best and highest-skilled team he'd already identified—but nearly every single step thereafter. Chris was thinking in terms of years. He also knew who was on his team. And, most importantly, he had a firm grasp of the problem he was trying to solve.

"I'd had my share of legacy issues with banks," Michel says. "I've seen it all. All the legacy-systems issues, problems with IT, and problems with the back end."

Michel got it. And Chris knew this as well.

"Most investors I was talking to had no idea what I was talking about, but Michel totally understood it within the first couple of minutes," Chris acknowledges.

"I thought, *If Chris could pull it off and could do what he is thinking—create a factory where other banks can hook on to and do their back office—that would be something that could become very valuable,"* Michel remembers.

THE VISIONARY IDEA

The *idea* was, in Michel's mind, brilliant. However, that's not what convinced him to invest in Chris. Rather, it was Chris *himself* that sealed the deal.

"I looked at this guy, and his idea resonated truly with how I thought it should work and where the future of banks—the finance industry—was headed. That resonated *deeply*. But then this guy, he was not just sitting there. He wasn't all talk. Believe me, I had other guys also talk to me about creating new banks and new payment systems. But Chris had a plan. He had a team. And what I then saw was that he had a group of people around him who had the skill sets that were totally unique. Most importantly, he had a *vision*."

This was a game changer for Michel. He knew Chris could see what others could not or did not have the balls to do. Perhaps he recognized a bit of himself in him. He also could see that Chris had a crystal-clear direction and intuition that guided him in a way that others simply didn't have.

Michel was also impressed by the team Chris had already gathered around him: Erik, who really understood a bank from a product and process perspective; Ilco, who could actually code the software; and Bas, who knew infrastructure and how to put it all on the cloud. On top of this, Michel appreciated that they each had a proven track record—they were all there at the beginning of BinckBank, and they had all helped it grow from a start-up into the international financial powerhouse it had become in less than a decade. "While Chris was talking, I was sitting there thinking, 'Those guys built a bank; they created a company from scratch with half a million customers, took it public, and it's now valued at one billion euros. These guys already know how to do it. They're really focused. They can sit down tomorrow and start building if I wrote a check out today,'" Michel recalls.

IF YOU PAY PEANUTS, YOU GET MONKEYS

Chris was honest with Michel. He had another interested investor in England who he had been talking to but who wanted Chris to ante up $25 million to make sure Chris had some "skin in the game." Chris understood, but this wasn't the ideal scenario for him. He had made it clear to both investors that he needed the €3 million up front. "I needed to pay all these guys—to see the project through and have a core banking engine we could sell," Chris says. Chris also explained that he wanted everyone to be 100 percent focused for the next three years. He did not want them to worry about their mortgages or have to take out loans. He wanted them to have a good salary, a share in the company, and that ability to focus. Michel understood this. Other investors were thinking only about the money and how much equity they could get in the company, but Michel was thinking more like Chris: *How can we get this done?* Chris adds, "Michel understood my vision of creating the core banking engine and keeping a staff happy and focused. Like me, he understood that all these people I needed down the road all had very good jobs *now*, and so I had to make the job and salary that I was offering them even more appealing. As they say, *if you pay peanuts, you get monkeys*. You have to remember this was also in the middle of the biggest financial crisis ever. People with good jobs were not going anywhere."

Though all the guys knew what kind of person he was, Chris still felt he needed to prove that he could pay them for three years. That is why he requested all the money up front—and with no milestones. Michel respected Chris's desire to provide for his team and respected his rationale and was happy to comply. Michel adds, "I would want to bet on these guys. With the drive they had, I knew they would pull it off. They would give it their all."

When Michel told a friend about the concept, the friend told Michel, "It's never going to work. The team will just end up keeping rebuilding the platform, and the regulators will never allow a bank platform to sit on the cloud." Michel recalls this conversation with a wry smile. "But they proved him—everyone—wrong."

Like Erik, Bas, Ilco, and Chris, Michel had something to prove, too. "I thought if they can pull this off and make this work, the future will be different than it is today."

Michel wanted them to succeed. He was all in—no strings attached. "A lot of these investors try to grab guys like these and tie them down and make them sign all kinds of financial stuff saying that if they walk away, they have to pay it all back. I knew I didn't have to do that with this group. I thought, 'We're not going to do that at all. We're just giving them the opportunity to do this, and when it's successful, we'll profit.'"

Michel, of course, knew the risks too. He knew the biggest challenge would not be for the team to build the platform—it would be changing the minds of those who run banks. It would essentially call for financial institutions to change *how* they do business. If they could see what he did—that SaaS core banking engine on the cloud was the future, the *only* future—then that future would be wide open. There were, according to Michel, "so many fish in the sea." The opportunity was limitless.

Michel, forever the optimist, was undeterred, and he knew that if anyone could change the minds of financial institutions, take on the regulators, and build a bank from scratch, it was the confident, slightly rebellious, long-haired, tattooed, and extremely determined man sitting before him. He was also optimistic about the banks as well. As one who worked in them, he knew there were plenty of guys like him who were also looking for ways to improve processes, cut costs, and be more efficient and secure at the same time. "There are so many banks and financial institutions that are changing and *do* want to change. That is a huge opportunity."

THINGS ARE FINALLY LOOKING UP

Michel didn't take long to decide. However, on the day Chris was due to arrive in his office and sign all the final paperwork, Chris and Myrthe were in the midst of a family trauma. Myrthe's father had a tear in his aorta and had been in a coma for ten days. The doctors had given him only a 1 percent chance of survival. For weeks, Chris had been not only negotiating the contract with Michel and the lawyers but spending several hours a day both at the hospital and at Myrthe's parents' home trying to handle the family's needs. Chris asked Myrthe to just give him some time away so he could complete the deal. "Please leave me alone for just a couple of hours."

When Chris arrived at Amerborgh and sat down to sign the final agreement, he received a call from Myrthe. He apologized to Michel, Alex, and everyone at the table, but he explained why he had to take the call. Myrthe's father's condition was precarious. The doctors could not revive him from the coma he was in, and every day that he remained in the coma, the greater the chance he had of dying or having lasting physical damage. The family was in a wait-and-see mode, but Myrthe's mom was preparing for the worst in case he might die. As soon as Chris answered, Myrthe delivered the news: "Daddy woke up!" Chris, overwhelmed with emotion, began to cry.

He didn't care where he was or who he was around. The past few weeks had been some of the most stressful days of his life. Yet in one day everything had turned around for him, for Myrthe, for Myrthe's father, and now for his business. He was overcome with joy. Life was looking up.

After all the contracts were signed, Michel wired the full amount over to Chris. Chris, at thirty-three years old, was now in possession of €3 million, the trust of an investor, and an idea he was certain was about to change the way banks do business. He first took a picture of the bank statement and sent it electronically to Ilco, Erik, and Bas, who all quit their jobs the second they saw it for themselves. Now he was ready to get to work.

They had a core banking engine to build.

PART TWO

NO COMING
UP FOR AIR
(2009—2012)

CHAPTER 4

BUILDING THE CORE TEAM AND GETTING STARTED

WITH THE MONEY IN THE BANK, Chris was finally ready to pursue his big idea. The first thing he needed was a name. When they started the company, they had some creative guys help them come up with their initial company name, Eeze. Chris looks back on the name and says, "We wanted it to be easy, but no one could spell it right. Good idea, bad execution." They also tossed around the name Kazaai but eventually passed on it because Kazaa was an illegal download site. One creative agency suggested Open—without the *h*—because what Chris was trying to do was be as transparent as possible in the financial industry, which up until then had been very closed. Chris thought, "It will be an open company. We have an open database, where financial software is very closed. The financial software world is closed, and we will be open."

But then Chris found out that Open.com was owned by American Express, and he knew they would never get that name or URL. So somebody, he can't remember who, suggested "put the letter *h* in between!" The word would still read as and sound like *open*. In short order, however, everyone started calling Ohpen "Ophen" (with a *ph*), inverting the letters when they read it. Nevertheless, Chris loved the idea and ran with it. The creative agency listened to Chris talk about the books he read on Buddhism and yoga and presented him a branding strategy and a logo—the ensō sign, a Buddhist symbol for togetherness, oneness, and openness—a simple circle.

Chris wasn't too worried about the name or its misuse. "All the big brands have odd names: Nike, Adidas, Google. Who heard those names for the first time and knew how to pronounce them or what they meant?" He was much more concerned about the brand. I wanted to make sure we stood out and were different from Temenos, Avaloq, Sopra, Misys, Oracle, Infosys, all these big core banking software players." After all, he thought, these names don't exactly roll off the tongue or stick in one's memory.

At its core, Ohpen headed by Chris and his fellow cofounders was going to reprogram the software that ran retail banks. They were going to make it faster, safer, and more reliable than it had ever been before. And, most importantly, they wanted to create it in such a way that there was only one version of everything—one codebase, one version, one API, one fully redundant infrastructure—that could serve multiple clients at the same time. This is what Chris dubbed "the Power of One." The reason he wanted to create one powerful core banking engine was fairly simple too. He and his team at BinckBank had learned the hard way how difficult it was to upload a new version of software and try to adapt it to a company's unique needs. "Working at BinckBank taught us everything that could go wrong," Chris says. "Changing, maintaining, and upgrading software and hardware were usually both risky and costly for banks. At Binck, the company we bought our software from, Syntel, was really bad at it. It sucked all the time. When we would get a new version, there was always something that did not work—always."

However, if he and his team could create a siloed multitenant platform built with the internet and apps in mind that would just always work, and every client used the same version of the platform (same version of the software) but had their own database sitting on that platform, he would essentially slash risk, costs, and downtime in one fell swoop. On top of that, by putting the software in the cloud, AWS, it would make everything better, the system would be reliable because it would always be up, and it would be superfast because Ohpen had new software and unlimited computing power. With servers with as much computing capacity as AWS, the Ohpen platform would have infinite uptime, performance, and scalability. If the team could pull it off and do what he had hoped, he would ideally have a system that always works, is always up, and is lightning fast, because they could cut down the processing time for certain main and core processes from eight hours to eight minutes. "That's a big fucking difference and a great way to sell it. We knew which processes were the most important at any retail bank, and we knew how much they cost to run and what they cost on a yearly basis, and now we could program them from scratch, loosely coupled with the other processes, with unlimited computing power, just as it was my idea at BinckBank many years ago," Chris says.

THE PIZZA TEAM

However, what Chris was asking of his small team was, from an outsider's perspective, virtually impossible. Banks with thousands of IT personnel at their disposal and software companies with thousands of developers working around the clock had not been able to do what Chris was setting out to do. And he wanted to do it with four people. Talk about doing more with less. Chris recalls why he wanted to keep the team so lean: "I liked the idea of how Amazon got started—using the 'pizza-team rule.' Bezos believed a team had to be just big enough to be able to eat two pizzas, but anything more than that would be too much."

Not long after Chris called everyone and said, "Dudes! Here it is! I can pay you!" and they all quit their jobs to join him, Erik found a small basement office for them on one of the canals of Amsterdam, Keizersgracht.

On their first day in the office, Bas purchased several desks and computers from IKEA and put them together. Immediately, Erik, Bas, Ilco, and Chris began working around the clock, which would be their practice for nearly five years straight. They knew they had a lot of ground to cover.

Ilco did the programming, Bas handled the IT infrastructure, Erik worked on the product, and Chris did the rest. "I loved the beginning, where Erik and I would be in front of the whiteboard for weeks, rethinking all these banking processes," Chris recalls.

A NEW LEADERSHIP STYLE AND THE BEGINNING OF THE OHPEN CULTURE

While Chris was abroad, he had ample time to do some soul-searching and made yoga a regular part of his life. Though he had practiced yoga from time to time during his years at BinckBank (and a bit more so when he was having back trouble from weightlifting), it was during his year off that practicing yoga became a routine. While he was in Asia for three months, he practiced every day. "I realized that I became calmer, and my fuse was getting longer," Chris recalls. "I still had fire in me, but it took a lot longer for me to get agitated. During that year off, it became a part of life." Soon after he made yoga a consistent practice, Chris began to meditate daily as well. Doing so helped him, he says, "become calmer, clearer, and more conscious." An added bonus was that his vision didn't, as he says, "blur," and he could clearly see what decisions he had to make. "The answers just came to me while meditating."

Through practicing yoga and meditation in order to stay more calm and present, Chris had arrived at a more conscious and mature level of being and felt well poised to lead the new company. He felt more centered and sure of himself—and his decisions—than ever before (though in truth he'd never doubted his own capabilities). Most importantly, he had built a team around him that he trusted and who he knew could deliver. "Those were happy days," Erik remembers fondly. Everyone got along well. And they felt inspired by Chris's enthusiasm as well as his belief in them. Each felt that what they were doing mattered. They also felt appreciated and valued, and they all were confident they were bringing their own level of core competence to the team. Each one of them was essential, and each valued the others. This was important to Chris and an integral part of his leadership style.

His own mentor, Kalo, had been tough on him. One day, when Chris was really happy with one of his proposals, Kalo put his proposal in the shredder in front of him. Chris thought that was tough—so tough, in fact, that Chris never felt appreciated or valued or heard the words *job well done*. Chris made a vow to empower everyone he hired and was determined to make them believe in themselves. He also vowed never to put anyone down. Rather, he wanted to encourage employees and motivate them with positive reinforcement. "Saying *job well done* and *thank you* goes a long way," Chris adds. From day one, Chris wanted to make all of his cofounders and whomever he employed feel valued and believe that their job—no matter what job it was—mattered to him and to the entire company. "But respect works both ways," Chris attests, "so they should respect the company as well." Chris admits he didn't always get leadership right. "It was a learning-by-failing experience for me," he says. "Sometimes stress got the better of me, and I did the opposite of what I wanted to do. Yet over time I started to get the hang of it."

> **Chris made a vow to empower everyone he hired and was determined to make them believe in themselves.**

Not long into starting the company, Chris realized he needed assistance. Erik, Bas, and Ilco each had defined tasks. Erik was working on designing the processes, Bas handled the infrastructure, and Ilco did all the coding, but Chris had to run and manage everyone; scope clients; build a proof of concept; manage marketing, sales, and PR to sell the product; run all of human resources (payroll, benefits, etc.); update Michel; and keep everyone fed. It was, in a word, a lot for one person.

CHRIS HIRES EMPLOYEE NUMBER ONE

Lydia van de Voort had just become Chris's neighbor and was looking for work. Lydia, a towering six-foot-tall gorgeous neighbor with long and wavy hair worthy of a shampoo commercial, was just twenty-four years old when she met Chris. She had recently moved into his apartment building after finding herself without a home. The internship at KLM she was to start in New Delhi, India, had fallen through just after she gave up the lease on her apartment. However, her father, who held a pied-à-terre in the same building as Chris, offered the space to Lydia until she figured something out. She had recently left school and didn't know many people in the building, so her father encouraged her to attend the community owners' meeting so that she could meet some people. At the meeting, when everyone was introducing themselves to each other, Lydia explained her circumstances and found herself saying aloud, "What do I do with my life now?" Chris had an idea: "I'm looking for an intern." Lydia, now looking back, can't believe her response to Chris, which was: "Okay, let me think about it."

"Let me think about it," Lydia repeats incredulously, retelling the story. Lydia says with a laugh, "If someone would say that to me today, I would say, 'Well, if you don't want to, you don't have to!'"

Rethinking her initial response, Lydia followed up with a text right away, asking to meet up. Chris texted her back, suggesting she join him and the team for some coffee the following day. She took down the address and headed over to the Keizersgracht to meet him and the team. When she arrived, she saw what most did: an unmarked door that led to a souterrain. Though the building was in a historic and pretty spot—right on a beautiful tree-lined canal—the large steel bars over all the windows seemed menacing and the space impossibly small for an office. When she knocked on the door and then peeked through the windows below, she saw that the guys were so engrossed in their work in front of their computers that they barely moved. She called through the door and announced that she was looking for Chris. Someone got up, let her in, told her Chris was running late, and returned to his work. She stood there for several moments before Erik, who had earphones on, finally noticed her, pulled his earphones off, and asked her if she wanted any coffee. She said yes, and without fanfare or so much as a welcome, he got up, poured her a coffee, handed it to her abruptly, went back to his computer, and then—without a word—put his earphones back on. Much like Matthijs's first experience at BinckBank, Lydia wasn't impressed by the guys' hospitality. She thought to herself, *What is this place? A bunch of dudes sitting around computers who don't talk to anybody?*

After some time, Chris pulled up outside the basement in his scooter, and Lydia observed him in a new light: not as her neighbor and father's friend but as a potential boss. "He had long hair combed back like a smooth Amsterdam guy," she remembers with a smile. She had come prepared and had printed out her résumé. When she handed it to him, Chris didn't bother looking at it. He wasn't interested in the paper before him at all; rather, Lydia adds, "He was more interested in the personality that he was talking to. Something that I picked up from him and use to this day. Though I always look at the résumé, I don't pinpoint things or focus on them. I'm always interested in the person who is sitting across the table more than what's on paper." Before she realized it, Chris said, "Well, when can you start?" Lydia responded, "Tomorrow or so?" He was taken aback by her willingness and said, "Well, that's too soon, because we need a computer and to get you set up, but why don't you come back in a week?" Lydia did just that. When she arrived for work the next week, she became Ohpen's first official employee.

When asked why he hired Lydia—after all, she had no experience and didn't know anything about computing or the financial industry—Chris said he liked what he heard when Lydia was speaking.

Lydia didn't have to work. She was the daughter of a successful entrepreneur. She could have been out enjoying life with her friends, but she *wanted* to work. That, to Chris, meant more to him than anything. But it was also the bare minimum. The guys he worked with had already proven their mettle time and time again at BinckBank, but Lydia was fresh. He wanted to see if she had what it took to be part of a start-up.

"When we started off, he really tested me," Lydia remembers. "He tested me to the point that I thought, *This is just impossible. He is challenging me. He is confronting me. He is pissing me off. He is doing so many different things and constantly trying to get me out of my comfort zone, only to put me into my next out-of-comfort zone.*"

When she started, Chris wanted to see just how much she would be willing to do. He was also, he admits, testing her ego—to see if she felt she was above anything (something he often does with new hires). He even sent her out to do his grocery shopping one day.

"Grocery shopping? You want me to get your groceries?" Lydia remembers asking him twice.

That is what he wanted, so that is what she did. She also got the guys meals and made them lunches. She even cleaned out Chris's car after one of his cats had become ill. Nothing, *absolutely nothing*, was beneath her. That was, however, until she reached her breaking point. When the team eventually needed more space to grow and moved from the Keizersgracht to the Herengracht into an old bank, Chris asked Lydia to sort through the rubble in the basement of the new office—which included hundreds of safe-deposit boxes, emptied and piled up in the middle of a room along with hundreds of fallen bricks and stones and out-of-place keys. Lydia's task was to find and organize all the keys that were strewn about and match them to the drawers that were spread all over the ground. "It was a disaster zone," she recalls. Lydia worked tirelessly—even cutting and bruising her hands to sort through the rubble. Her blood, sweat, and tears, quite literally, were poured into the foundation of Ohpen. There was no task too small—or even too large—for her to undertake.

"He definitely pushed me to my limits," she says with a laugh.

However, the safe-deposit-box incident was for her, she recalls, a turning point in their relationship. She had finally found the boundary within herself that she didn't want Chris to cross. She went to Chris and said, "This is just not what I expected to do! I'm an intelligent woman! I want to develop myself, and that means not just pushing away some bricks and tearing open my hands and numbering keys!" Then she said, "I'm done. I'm done!" Lydia can't recall the specifics, but she thinks she may have cried a bit for the first time at work. "I was just fed up and so frustrated."

From that moment on, Lydia believes Chris started to see her in a new light. *She was tough. She could stand up for herself.* Chris admires fighters. Despite their different pasts, he may have even seen a bit of himself inside her dark, fierce eyes. Though Lydia grew up in a loving and, by most accounts, privileged home, she had been taught to pitch in from an early age, and she was told there were no free rides in life. Like Chris and the guys, she came to work every day feeling like she had something to prove—to her family, to herself, but mostly to them. But unlike the guys, who felt like they had to justify not being college graduates, what drove Lydia was something else altogether. "I think I've always had that *I will prove you wrong* feeling inside me. I think that's a feeling that all females have—whether it's because of our environment or industry—I am not sure. But we have to constantly prove we're worthy." Like Matthijs and Chris, Lydia felt her competitive spirit come alive for the first time on the hockey field. "I had always been—and I still am—a harmonious person. I don't like to start fights. I was never the 'tough' girl. But I played field hockey quite seriously, and my parents always told me that there was one Lydia *on* the hockey field and one *off* the hockey field." Lydia played to win. That competitive edge drove her in the early days of Ohpen. No matter what the task, she says, "I came to get it done."

In the early days, what they counted on her to do varied. Eventually it transitioned into meaningful work for her. From the beginning, Chris included her in all of their meetings. It was truly an "open" environment. There were no offices or desk dividers, and everything was spoken about and discussed in the wide open. If Chris was meeting with someone, he'd take Lydia with him so she could take notes—transcribing every word and organizing them afterward. This gave her unique insights into how Chris thought, how he handled people and situations, and what it was he was trying to accomplish. She worked closely beside Chris and quickly learned the ins and outs of his daily routine and just the huge scope of work that he was undertaking. When they were working on process flows, Lydia observed and picked it up quickly. At one point, the team was building their website and needed to have a flow for the change-password item. Chris turned to her and said, "Can you work out this process flow, please?" She said yes, though not even knowing what she was getting into, and did what he asked. To Chris's (and her own) surprise, she nailed it. "This is actually quite good," Chris said, "so maybe you can make a few more."

Lydia was more than happy to oblige. "I think that, at least for myself, I felt I could contribute to the work the guys were doing."

Looking back, she can hardly believe what they all accomplished in those early years—and how much they have all grown since, both professionally and personally. "It's been a ride and it's been a challenge, but in every possible good way. I've always had many conversations with Chris since the day I started. We both feel that a person's professional development and personal development are always so closely related. From the beginning, he's been a real mentor to me in terms of those two developments, but it was never an easy ride. Because if personal and professional development were so easy, then why wouldn't everybody do it?"

Eventually, Lydia took on more responsibility. As Chris started to sell the software and go through RFP processes, she began to help Chris with contracts. When Ohpen wanted to grow through acquisitions, Chris began looking into buying a company, and she worked on the due diligence process. She applied to get funding from the state for innovation as well. She worked mostly under the radar. Chris says, "She did things that were really important, but a lot of people didn't see it or even know about these things. Every contract we had with any supplier, Lydia and I did together. When we did the first contract with Robeco in 2011, she, with only two years' experience in the company, was at the table with me to negotiate the contract. She knows *it all*: exit for convenience, exit for cause, service-level agreements, liability, limited, and unlimited. I would give her an NDA and ask her, 'Can we sign this or not?' She did not have any legal experience—or any experience, for that matter—but would read through the entire contract and come back to me and say,

'Should it not work both ways, this confidentiality clause?' I knew right there that she was smart."

"I've always been kind of the right-hand woman for Chris—especially for the first seven years. I just moved in the direction that he moved in," Lydia says. "If he was working on the more commercial organizations and making sure that we got PR, then I was there to support him, work out the conversation, and make sure it was being followed up on."

JAN-WILLEM KOELEMIJ JOINS THE TEAM

In the beginning, Chris admits, hiring was all about intuition. There were no official job descriptions, no "cultural fit" discussions, no formal tests, no formal training, no mission briefings—all things that Ohpen does now when onboarding new hires. Rather, Chris was following his gut. "Culture is top down, never bottom up. So the culture of Ohpen is like a personification of Bas, Ilco, Erik, and me, because we hired people that we liked, that we liked hanging around with, and who had the same values that we did," Chris says.

Shortly after Lydia joined, she introduced her childhood friend whom she had known since kindergarten—Jan-Willem Koelemij. He had just finished his bachelor's degree and would be home for the summer of 2009. Lydia gave him Chris's number, and Jan-Willem reached out. When he showed up for the interview at the Keizersgracht office, there was no place to conduct an interview, so Chris and Erik took him out and sat in a garden at the hotel next door. Jan-Willem felt like he was literally in the "hot seat." Jan-Willem remembers sweating in the blazing sun while Erik and Chris looked so cool and relaxed across from him in the shade. Again, as with Lydia, Chris didn't focus on Jan-Willem's résumé or class rank. Rather, he was interested in him as a person. Eventually, the three men connected on sports. Jan-Willem played semiprofessional football and was, by his own definition, extremely competitive. He was also up front and said he could only work for the summer because he was returning to school for his master's degree. Chris and Erik liked what they heard and took him on, throwing Jan-Willem several tasks a day, including but not limited to helping Lydia make lunches. *Nothing* was beneath Jan-Willem either. Menial labor and long hours didn't dissuade him. He loved working with the team and was "all in" from the beginning. In 2010, after Jan-Willem finished his thesis, he returned, joining the Ohpen team full time.

Though not an IT professional by training (he studied finance in college), Jan-Willem was exposed to and immersed in the world of IT during those early days as a start-up, when everyone was tasked to chip in. Whatever needed to be done, Jan-Willem volunteered. Just as Chris, Matthijs, Bas, and Erik had learned through experience at BinckBank, Jan-Willem did the same at Ohpen. "From the first day, I worked on everything from IT to finance to every other 'department'—whatever needed to be done," Jan-Willem says. Eventually, Jan-Willem, like Lydia, helped a little bit everywhere, from getting coffee to writing a manual for the CRM system to testing new functionalities. They even gave him what Jan-Willem calls the "impossible job" of singlehandedly moving the office from the Keizersgracht to the Herengracht, using nothing more than the company car—a Prius. (The car was leased initially to make sure Ilco—the all-too-important coder—could get to and from work safely. Up until Chris leased the car for him, Ilco drove a scooter. Chris worried about Ilco's safety, because he would work very late hours and needed to go on the highway to get home on his scooter. It was not only illegal to do so; it was very dangerous. So Chris got Ilco the car and forced him to drive it. Chris wanted to ensure absolutely nothing happened to Ilco. Chris admits, though, "Ilco used it … *sometimes*.")

Jan-Willem had to disassemble all the furniture, haul it out of the Keizersgracht basement office, pack it into the Prius, move it, assemble it again, and return, then repeat the process several times. "I moved the whole office—all the chairs, all the computers, all the books, everything—with Ilco's Prius. I think I drove like a hundred times," he exaggerates with a laugh. "In all seriousness, I think I finished at four in the morning. That same day, when everyone arrived at the office, everything was there," Jan-Willem recalls, shaking his head, remembering the move *and the heat*. "It was thirty degrees centigrade (nearly ninety degrees Fahrenheit), too. Looking back, I can laugh about it, but at the time, all I could think about was how hot I was."

Jan-Willem knew it was one of Chris's tests. "For the first five to seven years, everything was a test," Jan-Willem says with a Cheshire-cat grin, as if knowing what Chris was up to from the get-go. "The cool thing about Ohpen and the company is that you get a lot of responsibility and a lot of freedom to act on your own, as long as you make arguably good decisions and can explain why you do things." Over the years, Chris has put Jan-Willem and his fellow employees into a lot of situations where he asked, "Okay, how would you solve this?" Or he would be more direct and say, "Solve it. Then tell me why you would do this." Jan-Willem thrived under these conditions. He knows it's not an environment for everyone, but it worked for him. "Freedom and responsibility are very important to me. I could fail a lot because I had this responsibility and freedom, but I could also learn a lot as well."

Another thing that Jan-Willem both valued and benefited from was the "open" Ohpen culture. "I definitely loved that the founders—all four of them—are very easy to approach and are very direct. That's really how Chris and the others are. Most importantly, they really follow their own path. They have their own ambition and mission and really believe in them." Jan-Willem related to them on several levels. "We have the same kind of mindset. If we do something, we do it with two hundred percent input. That's who we are. That's who I am." In those early days, Jan-Willem's learning curve was steep, and he made mistakes. But Chris never yelled or screamed or even threatened to fire him. "I reported to Chris directly for eight years. When he explains how we should have done something that we did wrong, we can either accept that and learn from it (which I did) or not—and there are consequences. However, it doesn't matter to Chris if you fail or if you do something wrong, as long as you learn from it and try to do it differently the next time. And that's what I always did. So I did my best to understand his explanation and work with that. Some people ignore him, and unless there's a good reason why they do that, they're shown the door."

Though in the early days everyone was working around the clock—most of them putting in anywhere from sixty to eighty hours of work, seven days a week—they were having fun too, Jan-Willem recalls. "Everyone there was incredibly competitive, and no matter what the task or time of day, everyone was competing—even throwing something into a garbage bin became a fun, friendly competition worth a wager." They formed a kind of brotherhood (Lydia among them) and became very close. They spent more time with each other than they did with their own families.

BUILDING THE PROOF OF CONCEPT—OHPEN.NL

Now that Chris had his team and their full support, he could focus on launching the proof of concept. He knew that if he was going to go out and sell this software, he needed to prove it worked. He knew that all banks, before signing on the dotted line, would want to know the following: *Who else is using it? What's its track record?* Chris knew this was going to happen, so even as Ilco and the team were working on developing the platform and workflows, Chris was hitting the pavement and knocking on doors. "I went to banks and even offered our services for free. I knew we would need a proof of concept."

Chris even said to one private bank in the Netherlands that he approached early on, "We'll put your whole private bank online for just ten thousand euro a month. We'll build the entire thing." The bank still said no, giving as the reason that "people don't want online banks." Chris knew right then and there they needed a proof of concept—and they would need to create one for themselves. They would build their own bank, name it, get their own clients, and show banks how it works. But nothing Chris or his team did/does was "just because." If they were going to build a proof of concept, they were going to make it a *successful and real* one as well, with full branding, top-notch marketing—a wealth management platform that *just works* all the time for all of their very real clients, whom he would personally go out and pursue. In other words, on top of building a software company, Chris was building a retail bank of his own from scratch.

Chris wanted to be different—from brand to concept and everything in between. This was something Chris learned from Kalo at BinckBank. "Better to have two out of ten find you brilliant and the rest think you suck. It's better than everyone thinking you're 'just okay.' We have to be out there. We can't win on trust or being well known—just as BinckBank couldn't win on these same things when going up against huge, well-known banks.

> Chris wanted to be different—from brand to concept and everything in between.

We had to be different from the outset—a bit like W. Chan Kim's and Renée Mauborgne's *Blue Ocean Strategy*, a business book I read about unexplored market areas. We were creating an entirely new market and a new brand that did not exist before. Liking the name or not was irrelevant. I wanted to be a different voice in the world of all the same kind of voices."

By early 2010, Chris had the investor, the office, the team, a name, and a brand; now he had to build a bank—from scratch—and prove that IT. JUST. WORKS. And with a team full of the right guys with the right skills, not to mention a formidable woman with fierce determination, among them—all of whom had something to prove themselves—they were finally ready to dive in deep. No coming up for air until they came up with a proof of concept *they could take to the bank.*

CHAPTER 5

BUILDING A BANK FROM SCRATCH

FROM THE MOMENT ALL FOUR MEN started working together, even prior to joining Chris at the office on the Keizersgracht, they put their heads together to execute the plan. Yet before they even took a deep dive into processes, they first determined the scope—what their bank would look like, what the product offering would be, and what they *wouldn't* offer (mortgages or lending or credit cards). Chris thought that a lot of features in existing core banking engine software were not needed in the world of tomorrow. "If it's not good for the consumer and they don't need it, why build it? We didn't want to build derivatives or collateral into our platform, for example." He didn't want to split his focus; rather, he preferred to concentrate on the important processes and make them the best they could be. "I'd say to banks in my sales pitch, 'Don't change the software to fit your processes; change your processes to fit the software.'" Chris admits, to this day, that this is one of the most difficult concepts to convince banks to rethink.

Their ultimate goal was to offer a unique product—one that no one in the banking industry had attempted before. In addition to an operating system that "just works," they would also offer something no other bank had ever done before. "We called it back then a *bank out of the box*," Erik recalls. Because each of the men (Bas, Ilco, Erik, and Chris) knew intimately the troubles big banks faced on a daily basis, specifically in operating different systems and multiple versions of software to execute the basic needs of the bank, they felt that if they offered a service no other bank or software company had, they would differentiate themselves in the market-

> Their ultimate goal was to offer a unique product—one that no one in the banking industry had attempted before.

place. They knew that as a start-up company they could not build an entire banking suite (including one with mortgages and credit cards) to compete with the likes of Temenos or Oracle. Ultimately, though, when Erik and Chris were working on the scope, they were able to define the entire scope of their platform using three basic guidelines: *reliability, adaptability,* and *compliance.* Chris remembers, "Erik and I started writing down the features we would need if we would start a bank. We wanted to create a bank that was good for consumers. We thought about what we would need and what we would not need— or functional and nonfunctional items. On top of that, we looked at the biggest problems we'd had at BinckBank and what we would want to have. We knew we wanted to eliminate the nonfunctional elements. We wanted to focus on *reliability,* which meant a hundred percent uptime and superfast processes. The second thing we wanted to focus on was *adaptability,* which meant building an API that our clients could use to get everything out of the system quickly and logically and without needing us or paying us. This was something new in 2009. And the third thing we wanted to do was be *compliant.* We would take care of all necessary legal upgrades without charging clients for a new version."

FROM SCOPING TO DEVELOPING PROCESSES

Once they determined the scope, they needed the processes. They acknowledged there were some processes they couldn't rethink— such as calculating accrued interest. (It's a complicated calculation, but it's the same all over the world.) Other processes that could make the calculations faster could change. Especially the ones that made sense in the age of the internet. So the first thing the team did was make flowcharts in order to distinguish which processes would need to change and which ones didn't. The team took apart the entire process from customer onboarding all the way through the life of a customer to the closing of an account.

This too went fairly quickly. "A big plus was that our programmers (Ilco at first and later on, as we hired more employees, like Andre Janmaat and Anton Zhelyazkov) knew exactly how all these processes work," Chris adds. It's not uncommon for a programmer to need someone to tell him how to interpret the specs, how to read the flowcharts, which tests to write, and explain the decision logs. Chris recalls Ilco saying, "Dudes, I got this one. I don't need anything." For some of the processes in the beginning, there was nothing on paper. Ilco or Andre could just build it. "Well, we have a lot of experience, obviously," Erik says with a smile. "So we started with onboarding, then depositing money, then investments, then events within investing, corporate actions, dividend payments, income payments, and so on. Then there were withdrawals and closing of accounts—whether by decease of a client or just ending the service. So we defined all those core building blocks."

After they defined actions, they deconstructed each process one by one. "Customer onboarding—how should it work?" Erik asks. "Well, we built out a flowchart and created that process." Then they moved on to step two, making a deposit, and repeated the process. "It's logical to think about them together. How would that work, and how would that integrate? So how would you be able to deposit into your 401K during the onboarding process? How would they logically interact with each other in a digital and online world?" Erik recalls thinking through the process as if it were yesterday. "Then we got this third building block and started to think about creating the optimal process while cross-checking for consistency with an audit."

It all sounds easier said than done, and even Erik is quick to acknowledge this. "You have to imagine—we're talking about 2009 here. It was virtually impossible to open an investment account as a fully online experience as you would a social media account like those of Facebook or Instagram today. Logging on to Myspace and LinkedIn back then to check your account was normal, but for banks, to open accounts and make deposits online was actually unheard of. We were the first in the world to enable full customer onboarding online," Erik acknowledges. "I am pretty proud of that."

In addition, the team faced another complex hurdle to jump. "Corporate actions are incredibly complicated. There are basically thirteen or fourteen different types of corporate actions. You have your dividends, stock splits, reverse splits, liquidation, and more. All the other core banking systems use different transaction types for all these corporate actions," Erik adds. The job that lay before Erik and the team was to break down each type of corporate action and streamline the process so that it would not only be easy to maintain but, most importantly, perform all the calculations perfectly. And inherent in this was a major roadblock. Once they came up with the best, most streamlined, perfect calculation process, they had to *convince* banks that Ohpen's way was the best way. "Banks are very clingy to the past and what they know. Trying to be different is so goddamn ... excuse the word ..." Erik admits, shaking his head with frustration, "difficult." He acknowledges that while many of the people he and Chris were trying to convince in the beginning to use their product were intelligent, some simply couldn't wrap their minds around what Ohpen was trying to convey. The banks simply wanted "business as usual," and they wanted Ohpen to do things *their* way.

> Once they came up with the best, most streamlined, perfect calculation process, they had to *convince* banks that Ohpen's way was the best way.

"I think a big part of why we are successful today is that Chris said no to a lot of stuff early on," Erik says as a point of pride. "We ran a bank. We were looking at it from the inside. We were our own customer for ten years. We knew the way processes should be built," Erik adds. He also acknowledges the satisfaction they had in proving this. "What was really fun was holding up a mirror to them (bankers in the C-suite) and asking questions their managers didn't ask." The example Erik uses is trying to find the owner of an unspecified hundred-dollar deposit that comes through a bank. "You have to figure out for which customer it is, and you have to ask yourself, 'How long are you going to look for the customer?'" Most managers would say, "Until we find the right customer to whom that deposit belongs." And Erik's next question would be, "Well, how long does it take?" If you are looking for one person *for one day* to whom that hundred dollars belongs, that customer will *never, ever be profitable again in his whole life cycle.* The cost of man-hours to find the owner would be in the hundreds of thousands. "What the bank should have done is automate the return of the deposit," Erik says. It could have been easily built into the processes. If the customer still wants to invest, they'll resend it with the right description and with the right account number. In most banks, however, they are so big that they don't know what is costing them money. Chris sometimes asks C-suite-level executives, "When was the last time you visited the back office, or when is the last time you have answered a phone for customer services?" The answer is always, "Never." Chris explains, "A lot of banks just accept that certain costs are the way they are. With our first client, the total cost went down forty percent after the implementation of our system. And we saved another client thirty-one percent in total IT and ops costs. That is enormous savings." In addition to wasted expenses, banks don't know how much time is being wasted because processes aren't just automated—they aren't even thought through properly in the first place. From the beginning, Chris was trying to do just that—think things through in a way that had never been done before—to challenge mindset and challenge the concept of business as usual.

While some processes were fairly complex and difficult to convince banks to change to, other processes were fairly simple—but no less problematic when it came time to change the way banks do business. One of these processes was paperless processing. These days it's common practice—paperless statements, onboarding, billing, and receipts. But in 2009, everything was still being processed on paper in banks. The team wanted to create a fully integrated online experience—and that meant paperless statements, paperless every-thing. That meant everything would be backed up in the cloud. Everything would be accessible 24/7. They could save banks a lot of money—and the environment lots of trees and back offices lots of time (no batching, printing, etc.)—by streamlining the process on the back end.

"Thinking about how stuff works in the back end and how to make it really smooth and quick and fast," Erik recalls, "was fun." For Chris and Erik, all of this made total sense. They also knew what they were up against. Erik recalls one bank director who refused to use email—even in 2009. "He had his secretary print all the emails in hard copy and put them on his desk, and he would write a response by hand on the letter and she would type it in Outlook again and send it," Erik says incredulously. "No one was looking over this bank director's shoulder and assessing what it cost the bank—and by extension its customers—for him to print those emails, spend time handwriting responses, only to have someone replicate that process by typing up the response." At every level there were inefficiencies. Going paperless—just by using email or secure online document repositories—already cut those inefficiencies in half. And going paperless with bank statements and automated features could even further cut back costs. In one particular project, Chris recalls a client insisting on the ability to print statements. They believed they would lose clients if they had to download a PDF and print it themselves. "Make this feature, or we will not continue with you," they demanded. Chris said, "Have your lawyer send me this statement in writing, and then I'll make it for free." Chris never received the statement. Chris admits that if the bank really would have wanted it, he would have built it. "We were just trying to help. If banks want something different, we build it—as long as it does not screw up the core of our system and compromise our principles." Chris realizes that this tough approach sometimes rubs clients the wrong way, and he admits, "We can come across as cocky or wiseasses." However, his reasons for pushing back on clients who request the status quo are sincere. "We *were* our clients for ten years. We know what happens at a bank on a daily basis, and we know what we can do to fix things."

While some clients heard "You are doing your job wrong" and felt that Chris and the team were criticizing them, Chris had nothing but good intentions and wanted to help. "We had to adjust our tone," Chris concedes.

In addition to thinking about processes in a nuanced way to cut costs and save time, Chris and the team also wanted to think through the *level* of performance. They all knew the hardware wasn't fast enough for the software running on it—and the cloud would fix that. But, Erik adds, the hardware wasn't the only problem. "From my point of view, if you have performance issues, usually it's a software problem." Erik, Chris, and Ilco knew that if they were going to rebuild a bank's core banking engine, they needed to rebuild it all—from scratch. They had to make sure the entire thing was built from the ground up with the best coding possible.

DEVELOPING A BANK FROM SCRATCH—THE POWER OF ONE

When it came to coding the bank, Ilco "Finish What You Start" van Bolhuis was the perfect person for the job. "I think what I do differently than most is I make it work. Always. My software works. If I think it's a bad design, it's not functional, overall not right, then no, I'm not going to do it. If it's a bad idea from the start, it's difficult to make it better or improve it," Ilco argues. "My knowledge of how I think an end user will use it is also on the table, and then together we try to make the best product out of it," he adds.

Erik, Chris, and Ilco knew that if they were going to rebuild a bank's core banking engine, they needed to rebuild it all—from scratch.

Ilco and other developers like him equate writing code to writing sentences—or a book. There are infinite ways to arrange words to convey a thought, message, or story. Similarly, there are infinite ways to create a code. And just as some sentences are better than others at communicating a message, some codes are written better and more efficiently and elegantly as well. Writing code, like writing books, takes advanced skill, talent, creativity, and experience. Ilco adds, "If you ask ten developers to write something in code, you get ten different implementations of exactly the same piece. There are so many ways to program stuff that everyone will do it differently, even if the outcome may be the same. But it may not be the most efficient." But unlike assessing writing (which can be subjective), assessing code is literally binary. Ilco adds, "The test is simple: *Does it work—yes or no?*"

When the team got started developing the software, they met each morning for about twenty minutes, and each talked about what they were going to execute that day. Ilco's job from the beginning was all about creating ultimate functionality. Ilco's first task was to build the software to talk to the database they had created during the scoping process. In addition to skill, Ilco had speed—thanks in part to all his years of experience at BinckBank. He was able to create databases in as little as ten minutes (just for some perspective, what Ilco was able to achieve could take some banks days, weeks—and in certain cases, because of all the bureaucracy and red tape—months to achieve). "But," Ilco demurs, "I knew that system. I knew what kind of tables I needed. So I knew I needed an account table, transaction tables, storage, locking, and testing." Ilco personally tested every single process, too, as he went along. In fact, at one point Bas did a software check and was running tests to see how far the software would go before it "broke"—or when it couldn't calculate any more. Bas ran the software on the AWS server and didn't uncheck the autoscale feature. So the program kept running and running—and could have gone on running infinitely. The program never broke. In essence, the software could go on in perpetuity, running calculations infinitely, at a massive scale. It would never break, pause, max out, or slow down. *IT. JUST. WORKS.* on a scale that none of the men had witnessed before in all their years of working for banks.

In addition to AWS's testing feature, *just to be on the safe side*, Ilco also wrote software to test his own stuff. "I wanted to create a shitload of random scenarios," Ilco explains, "completely *random* cases, just to test the functionality." He did so for one reason: "I wanted to be *absolutely* certain it worked. These are real assets and real money from real people. I had to be sure the numbers were correct." The son of a potato salesman did not forget his roots. He was and still is looking

Erik and Chris in the office the day the system went live at 5 a.m.

Chris took everyone skiing as an acknowledgment of their sacrifices.

Top: The team went skiing again after Robeco went live successfully.
Bottom: Very proud moment when Amazon named us as the first bank ever
to be put in the cloud.

Chris and Erik before signing the first deal with Amerborgh,
the first investor.

Lydia working all night.

First contract signed and sealed.

Ve worked 24/7 for three months on this first contract. We were so proud.

The new Amsterdam office in 2012.

out for everyone, knowing full well that at the end of the day, users are just trying to put food on the table for their families too.

All in all, it took Ilco anywhere between one and five days to build each process, though some processes required more time. In order to build the bank, he had to build a total of two hundred processes. By the end of many months, they had built a complete and working software system—a singular all-inclusive system. "We built everything in one system. The 'Power of One,'" Ilco says with a sense of satisfaction, knowing that he was the Power of One. "I was able to do it so quickly because the team was so tight and well managed." However, Chris soon realized he needed to hire others to help Ilco. Ilco admits, "I was very naive. I thought, *Well, I built it! I'm done now.*" However, Chris was out selling and completing an RFP process for their first client, Robeco, and the bank naturally wanted to incorporate capabilities and had further requests.

"It was just a shitload of stuff, and I was unable to do all of it in a short period of time on my own. I needed help, and Chris figured that out," Ilco adds.

Though there was one core banking engine, banks still had functionality and customization requests that needed to be processed. Though Ilco had worked with a pool of twenty-five developers while at BinckBank, he was willing to work with only four of them. "Those were, in my opinion, the best developers out there." Among those developers were Andre Janmaat and Anton Zhelyazkov, who still work for Ohpen to this day. Each man brought with him his own specialty. Some were good with databases, and some were good with API. "Anton's very good in API. I like databases more, and Andre is great at back end systems," Ilco adds.

From 2009 until 2012, the team worked day and night to make sure the program worked perfectly, slowly adding new team members as the workload increased. What began as a team of six—Chris, Erik, Bas, Ilco, Lydia, and Jan-Willem—began to grow slowly and steadily as they developed the core banking engine's basic functionality and proof of concept.

While Ilco was doing his best work and making sure the core banking engine worked perfectly, Chris, Erik, and Bas knew that the guys in the C-suite weren't worried about software functionality. In fact, they probably had no idea what Ilco was doing. "At the board level," Erik adds, "they just want to know, *Is this system up? Is it secure? Is it compliant from a regulatory point of view?*" Erik adds, "If I'm a board member, I am liable. If I'm not compliant or if I am hacked or I go down, I get in trouble. These are the three important things." The team built their software with all of this in mind: it will always work, the system can never go down, and it is built with compliance and safety in mind. Erik recalls that most bankers, when Chris first pitched the idea to them, didn't even seem to understand what he was talking about. No one could quite believe they could do it—especially with such a small core team.

> **The team built their software with all of this in mind: it will always work, the system can never go down, and it is built with compliance and safety in mind.**

PART THREE

THE BENDS
(2009—2012)

CHAPTER 6

IT JUST TAKES ONE

IN DECEMBER 2010, Michel and Chris presented the Ohpen platform at a bank with a life insurance company. Michel remembers somewhat incredulously, "And instead of agreeing to use the Ohpen platform, the bank asked *us, a small start-up,* if we wanted to *buy the entire insurance company from them.*" Michel remembers them saying, more or less, "Your plan is great. But don't you want to buy us? If you buy us, you can also do what you want and use your platform in the life insurance company."

Michel and Chris started looking into it. They thought it wouldn't just be an interesting investment; it would ultimately help Ohpen create some more size and put a large infrastructure around Ohpen, including risk management, compliance, and legal departments (all of which would help them position themselves better in front of prospective clients). In addition, it would also expand their balance sheet. After an intense negotiation process, Ohpen went for it. "We bought this life insurance company." Michel shakes his head with a smile. "We put it together and opened it under one holding company. So when we started talking to banks again, we now had this life insurance company. We had the Ohpen platform running it. And banks could finally see that not only did Ohpen's product work; the company had its own risk management, compliance, legal department, and a large balance sheet—some 750 million euros. We had size, and that was something very important, because it made us a credible partner for large financial institutions. And we could show them we knew what it was to work in a regulated world, as we were regulated ourselves."

ANGELIQUE SCHOUTEN JOINS THE TEAM

In the midst of all this transition, as Chris finalized the due diligence process for the purchase of the insurance company that would help expand their balance sheet, Angelique Schouten joined the team as a part of the deal. Angelique made the choice to leave the "golden cage" a couple of years earlier. She had a senior-level position at a bank and was guaranteed a great salary, pension, and benefits for life. However, she was unhappy and unfulfilled. "I wanted to make a difference." She made the decision to go to Robein, the insurance company Ohpen was looking to buy, knowing full well they would ultimately be purchased. "I didn't know *who* was going to buy them, though," Angelique recalls, but she made the leap because she was ready. She began working for Robein heading their marketing and communications team and was enjoying it when Ohpen bought them. "I was like the furniture. I came as part of the package deal," Angelique jokes.

Angelique and Chris quickly butted heads during the writing of the joint press statement announcing Ohpen's purchase of the company. Chris wanted her to make specific changes, and she pushed back. The two argued, and Angelique remembers a quasistalemate afterward. "We didn't talk for two months, and every time he heard my name, the two other Robein board members said that his neck would get red." Angelique laughs now at the memory. Chris has a different take. "It was Ohpen buying *them*, so I thought she was pushing back to make a point, not because her argument was the better one." Chris also felt there were better ways to make a point. "It was not that I did not talk to her for two months because of that. If I'd wanted her out, she would have been out at the end of the day." Chris, in his own estimation, had what he calls "bigger fish to fry." With the acquisition of Robein, his staff grew to over one hundred. Chris didn't have the bandwidth to fight with Angelique—or anyone else, for that matter. As far as he was concerned, he had a plan that he needed to execute. Chris adds, "I think it was a bigger thing for her than it was for me. I just thought she was wrong and overreacted, but I did not mind." He was, as always, more interested in performance. "We'll see later if she can do something or not," Chris recalls thinking. Eventually, though, the two became more collaborative, and both realized they were more alike than different, in many ways cut from the same cloth.

Like Chris, Angelique had had a tough background. "We both are not afraid to fight," Angelique says. When Ohpen acquired Robein, Angelique was originally put in charge of running the retail side of Robein and Ohpen's proof of concept asset manager, ohpen. nl. But in short order, everyone on the team quickly realized that if they were going after other banks like Robeco, they couldn't compete with them as a retail bank as well. Angelique recalls, "When we were in one of the best pitches in our history, one of the prospects asked us, 'How can I convince my board to go with Ohpen when your TV commercials are stealing my customers at the same time?'" Later, out in the parking lot of a prospect, Angelique turned to Michel and said, "Maybe we have to kill our own darling?" She knew that meant her own job was at risk, but she wasn't worried. "I always say that as long as you can say, *Hi, welcome to McDonald's. May I have your order, please?* there's a job for you." Angelique was not afraid of hard work and made peace with the fact that Chris may let her go. She was also aware that Chris was already thinking of "killing his own darling" as well. However, Chris recognized that Angelique was driven and had dedication, which aligned with Ohpen's core values. Chris recalls, "She was a bit of a diamond in the rough, and she needed to learn to pick her battles." This was just a small hurdle to him, and he had a sense she could overcome it. For him, the *work* was the most important thing, and he wanted to see if she could deliver. Chris says, "I don't care if someone is black, white, tall, short, young, old, woman, man, has ten degrees, or is self-taught—I only look at the execution of the work." In the end, he liked the work Angelique produced, so he gave her a shot. "After that," he adds, "it was on her."

Angelique had more in common with Chris than either knew at the time. She too grew up with an abusive father. When she was eleven years old, the abuse escalated, and she started to fight back. "I wanted to protect my mum. At one point, I packed his bags and put them out on the curb." Angelique's mother divorced her dad in the end, and Angelique, in her early teens, had to testify in court against him for not wanting to keep the same last name as him. Her father cleaned out all the bank accounts. Shortly after that, he moved to Thailand, remarried, and never paid child support, leaving Angelique, her sister, and her mother penniless. "My mum lost their company, lost everything, and she couldn't work." Like Chris, Angelique started to work at a young age. "If you want something in life, you work for it, and you give it your all. As long as you can talk, stand, breathe, you can work. No one's going to give you anything for nothing. That has been imprinted in my mind from a young age." As soon as she could work, Angelique knew she needed to find what she called "performance-based work" because, she adds, "I could work harder than anyone, and nothing could stop me." She started planting leeks. She literally crawled on her knees in the dirt, and every ten centimeters she planted a leek for 750 meters at a clip. "I was really fast. So, I made my first money." Eventually, Angelique went to university at her mother's urging. Her mother, who grew up in a large Catholic family, wasn't allowed to go to college like her eight brothers as "she was destined to marry," so she felt strongly that Angelique should go. Angelique funded her entire education on her own, taking out loans and working her way through it. But by the time she entered the workforce, she was €100,000 in debt. "I just started working like an idiot. Like crazy. And that's been constant throughout my career, and I think that's also the mentality I have. *Just do it.*" Chris saw this fire and work ethic in her and liked it. There is nothing that will get in Angelique's way once she puts her mind to it. And she's been committed to the company and its give-it-all mentality since day one. She adds, while recalling the past eight years, "It's been a helluva ride."

Like all the rest—Chris, Erik, Ilco, Bas, and Lydia—Angelique has felt that she has something to prove and shares that bond with all of them. Though she feels especially bonded to Lydia. "Being women, we have to prove ourselves ten times more and still more." She admits that tech, finance, and leadership positions are largely dominated by men, and it's difficult for women to prove that they have a space at the table. She adds, however, that Chris has always had her back in this area. "He doesn't care what race, religion, or gender you are. It's about doing the job—the outcome."

Lydia was not alone anymore; there was another female in the ranks. And there was plenty of work to go around for all of them.

A MAJOR RFP

Then, in early 2011, Michel tipped off Chris that there was an RFP he had heard about. Robeco had launched an RFP. This was the opportunity Chris had been hoping for. He and Ohpen could compete in all three categories, which had never been done before. At the time, a core banking SaaS was not known in the financial world. Erik, Lydia, and Chris prepared for a multiday pitch, knowing full well they were going up against core banking giants. "We were just a team of a couple of folks from Amsterdam, showing up at a massive building in Rotterdam," Erik remembers, shaking his head. They arrived wide eyed and awestruck as they entered the three-story atrium of the twenty-five-story bank building. "We felt a little bit like David and Goliath," Erik remembers. "It was really huge. And we felt really small. I think this actually helped us. We felt like underdogs and like we had nothing to lose." Chris and Erik were relaxed and confident—even making jokes—in front of the panel comprising the forty-plus people who sat at a large U-shaped table in an arena-type room. Despite Chris's and Erik's informal demeanor, the panel was ultimately impressed by the small-but-mighty team that stood before them. In fact, for every session and pitch, Erik and Chris showed up surprising the panel. *"Oh look, it's the Chris-and-Erik show again!"* Erik recalls, "I think that made a lot of impact. We were different from the other big names who sent in separate 'business development teams' to do each pitch and then who simply asked at the end of each presentation: 'Any questions?'"

In contrast, Erik and Chris took a more conversational approach. "We really engaged with the panel," Erik remembers. The other thing Chris and Erik had going for them was their track record— they knew how to build a bank. "We really were able to make them understand that we were once where they were. We experienced their problems firsthand as operational managers of a bank. And as we listened to them telling us what their issues were, we were able to say, 'Oh yeah, we had that back then as well, and this is how we solved it.' They knew we didn't only understand IT but really understood their business in the end. Obviously, because we ran a successful bank for many years."

While *Erik* may have felt like they had nothing to lose, Chris was feeling the heat. It had been two years without a client or a sale, and he was running out of capital. If they landed the Robeco contract, he would need to hire more staff. Chris had to call Michel and tell him that because of Robeco's size, he would need to spend more money and hire more of his old staff. He couldn't wait for Robeco to sign. "It will be too late. We have to do it now," Chris explained to Michel. Chris knew he was betting the entire company on this call. If he hired everyone he needed, maxed out his capital, and then didn't sign Robeco, he would run out of money in three months. Nevertheless, Michel believed in Chris. He said, "Do it. We'll back you." Chris never even showed Michel the budget. "Michel trusted me," Chris remembers. "I had a lot of stress until I knew Robeco would sign. But I did not tell anyone that if we lost Robeco, we would be bankrupt three months later."

The team did the best they could during the pitch, and now the decision was in Robeco's hands. Chris went back to business as usual—traveling all over, trying to sell, as well as building out the proof of concept and conducting due diligence on the purchase of the insurance company—in order to expand the balance sheet.

THE DAY IT ALL CHANGED

Chris, who was at Charles de Gaulle airport, received a call from Robeco. Having just arrived in Paris, Chris was walking on the jetway when he heard the news: "Hi, Chris. I'm calling you to say we want to continue with you guys." Chris was almost in disbelief. "I'm on the plane. I can't talk. I have all these people around me, but thanks a lot. I'll call you back later," Chris said, hanging up the phone. Chris wasn't being rude or dismissive. He just couldn't contain his emotions. "I was like, *finally*. After all these years of getting rejections, of fighting, it was over. For two and a half years, I got so many rejections. *It can fucking kill you. It can eat you inside.* I found it a very tough pill to swallow." And suddenly, there was nothing to fight against—nothing gnawing at him from the inside. *Robeco wanted to be Ohpen's first client.* Hearing these words struck Chris profoundly.

It was a mixture of exhaustion and elation. When he got off the jetway, he found a chair and had to sit down for a minute. He took a couple of deep breaths and let the realization settle within him. While sitting down, his eyes welled up. After years of struggle and unbelievably hard work, he felt completely overwhelmed and could barely move. Tears began to fall. He shook his head several times and stared off into space for some time, though he doesn't remember for how long. When he finally was able to get up and leave, he was almost stumbling around in disbelief. He walked through the airport surrounded by throngs of people, none of whom realized the enormity of what had just happened and how much the trajectory of his life had suddenly changed in an instant. Nevertheless, he had to stay composed. No one around him knew what he was feeling, and he could barely understand, let alone control, his emotions as he made his way out of the airport. "I had been walking on the edge of what I could do. But I knew I had three choices in life—give in, give up, or give it my all—and I chose the latter," Chris says. Overcome, he was all alone with no one to tell the good news to. But Chris couldn't take long to celebrate or relish the win. He knew the real work was just beginning. He now had to deliver the proof of concept for a promised savings core banking engine.

GETTING THE DNB, ROBECO, RABOBANK, AND AWS ON THE SAME PAGE

For the next several months in 2012, Robeco conducted due diligence of their own and made several site visits to Ohpen. Because they were still so small as a company, at one point Chris called his wife and asked her to bring some friends over to sit at desks so that it looked like the company was full of workers. In truth, they needed all the help they could get. All the team members were relieved to have

Matthijs on board now too. Because the savings features were the most important part of Robeco's proof of concept, Ohpen needed to build the functionality—and do so quickly. Ilco had told the team, "If you can put a plan together in Excel, I can program it." And with Matthijs's help, Erik was able to build a prototype in a single weekend from his kitchen table. Over the course of one weekend, the two turned on some music, opened up their laptops, and started building a savings bank. The built-in-the-kitchen savings product ran for several years. The basic principles of the internet savings accounts still operate on the same code to this day and was so outstanding that it helped Ohpen win the Accenture Innovation Award for total automation of all back-office functionalities in that same year.

While the team was enjoying success and building out the functionalities, Chris was dealing with an entirely different level of stress. While negotiating the contract for Robeco, a major issue arose that could have had the potential to end Ohpen before it started. In order to ultimately finalize the Robeco contract, the Dutch Central Bank had to approve the use of AWS as a 'material subcontractor' and AWS had to approve the DNB's right to audit. This was a highly complex, intricate, and sensitive issue. Without the DNB's approval, Robeco (or any bank) would not be able to migrate their data onto AWS. And AWS's major concern was that they didn't want any of their clients' rights being violated or have prying eyes on their intellectual properties. If the DNB could audit them, it would be like opening a Pandora's box for future inquiries, jeopardizing other clients' privacy. "It was 2011. Nobody knew where the AWS data centers were. You can't just *get in*. It's like Fort Knox. Even if you do know, you can't knock on the door. You can't even go to the places where the actual servers are. They take their clients' security very seriously," Chris explains. Needless to say, it was a tense negotiation process that kept

Lydia and Chris up at night. Robeco said that if they couldn't resolve the issue by the first of January of the next year, they would pull the plug and could potentially shut down the project. In sum, the entire future of Ohpen rested on getting AWS; the Dutch Central Bank; Robeco and its parent bank, Rabobank; and Ohpen agreeing on a contract, which included the DNB's right to audit. "I need to thank one person for all this, and that one person is Leni Boeren, the COO of Robeco back then. First of all, she was the first one who gave us a shot. Secondly, she gave me the time to arrange this with the central bank and AWS. Many other board members would not have taken the chance or would have pulled the plug months prior. But she hung in there and trusted me. I will never, ever forget that. Without her and her trust, Ohpen would not be a company today."

The basic issues were that when someone becomes a client at AWS, everybody gets the same contract. What Chris ultimately needed to do was to get AWS to change this standard contract language to allow the Dutch Central Bank to do an audit at the Amazon datacenters. "If we couldn't get that done, there would be no cloud bank, because the central banks in Europe would have stopped it," Chris says. This was an enormous undertaking. Chris wasn't just tasked with getting four different parties to agree on a contract; he was ultimately changing how banks do business. It was a bit like herding cats. Chris had to manage Robeco's

Chris wasn't just tasked with getting four different parties to agree on a contract; he was ultimately changing how banks do business.

internal and external lawyers, the DNB, AWS's lawyers and external lawyers, and the internal and external lawyers of Robeco's parent company, Rabobank. Chris's main task was to get *everybody* to agree on this audit sentence in the contract.

Bas was at an event and saw that Werner Vogels, the chief technology officer of Amazon, who happened to be Dutch, was there, and he called Chris and asked him to be there in thirty minutes. With a lot of people gathered around "the rock star of the cloud," Chris pushed everyone away, jumped in front of Werner, and asked him for help. Chris knew what was at stake for Amazon and told him as much. "If you want the financial services market, you will have to arrange this. I know we are a shitty little company, and if you're Amazon, you are more interested in getting Bank of America on the cloud. But whoever you're going to sign, every regulator in the world will want this audit right. If you don't have this audit right, you will not get one financial regulated company." In other words, there would be no Capital One Bank on the cloud if it weren't for Chris Zadeh and Robeco, who will go down in history as the first to put a bank in the cloud.

Amazon, of course, understood this. They knew if they wanted to be able to service the financial world, they would need to facilitate the regulators. So they told Chris they would help—but they had some stipulations. They wanted time—a warning from regulators. They only wanted to disclose data under a court order, and they wanted a certified auditor. "That's how it started," Chris remembers. "The DNB was like, 'Oh, wait, we can't have any restrictions.'" Chris then became the de facto mediator. He said, "Okay. But let's say I'm a cleaner/janitor—and there's nothing wrong with a cleaner—but don't you think Amazon has a point if they say, 'I don't want the cleaner to come in and see the data?' And the DNB said, 'Oh, of course not.' And I said, 'Shouldn't it be someone who knows what to look for? Say *a certified auditor?*' And then the DNB said, 'Yes, of course.'" And then Chris needed to wrangle all the lawyers. So the

AWS lawyer said, "It has to be a certified auditor, and *we* decide if it's a worthy certified auditor." But then the DNB said, "*We* decide if it's a good certified auditor." The different parties then began to argue about who would be the judge of said auditor's credentials. He and Michel often joked, "If you want to kill a deal, bring in the lawyers." Chris knew that with so many lawyers at the table, it would be a daunting task getting everyone on the same page. "For months, I was like this. *No, no, no, no. Don't leave! We'll get there. We'll get there. Don't leave. Don't leave. They have a point. No, no, no, no! Wait, wait, wait, wait, wait!*" Chris says, gesturing with his arms as if wrangling animals into a cage. Everything was on the line for Chris. But Chris wasn't alone. AWS was a great support. They knew that if they got the approval of the DNB, it would be the gateway for all European national banks to follow suit.

For nine months, Chris and Lydia pored over every version of the contract that moved between the four parties. "We just had one paragraph that needed to be okayed by everyone. But it took longer and longer." While the team was working on the proof of concept and meeting deliverables, the contract still held the processes back. If the DNB, AWS, and Robeco couldn't agree, Robeco threatened to pull out of the whole thing unless they came up with an alternative scenario. In September, Ohpen began working on a backup scenario.

"The climax was actually that after a while, we agreed on what should be in the contract. We agreed that AWS would need notice of the audit, unless there's a big financial crime, and then the audit could be done within twenty-four hours. There were all these rules. Then we agreed on the principle of the audit right clause. Then the lawyers started writing it down, and things started getting tense," Chris remembers.

After nine months of wrangling, things came to a head during the Christmas week. Robeco wanted everything finalized by January 1. If it wasn't, they would walk. Chris knew that everyone stopped working during the Christmas and New Year's holidays. "So it was now or never."

On the day before Christmas Eve, Myrthe and Chris headed out to their scheduled holiday in Zermatt, Switzerland. Chris was driving the ten-hour trip through the night when he got a call from Amazon saying they'd finalized the wording. "This is it: the last version. Take it or leave it." Myrthe was asleep next to him, and he knew he had to get Lydia on the line to go over the contract with him by phone. Chris knew *this was it*. He called Lydia, who was outside enjoying her second glass of wine while at a party with friends. "You have to get to the office now," Chris said. Lydia got on her bike wearing a short dress in the freezing cold and rode over to the new offices in Rokin, the whole time thinking, "Shit, shit, shit!" She knew she had to get clearheaded to do what she was about to do. All the lights were out when she arrived at the office. She ran in, turned on the computer, and got on the phone with Chris right away. "Let's do this." Chris then started making calls to all the parties—lawyers for Robeco, lawyers in Seattle, and the chief of Amazon in Holland, Kamini Aisola, "who was absolutely amazing in helping us get there," Chris recalls. "I was on the phone with each at different times, and then with Lydia, who was working on all the versions. It was midnight in Germany, where I was at the time on the road, and around three in the afternoon in Seattle."

Chris remained on the phone with Lydia throughout the night—some six or seven hours. Lydia kept going over each version, making sure each line was exact and collating all the changes from the various parties. Finally, they had all parties in agreement in the wee hours of the morning. "I'll never forget that moment—in the middle of a fucking German highway. I think it may be the thing I am proudest of. More than building Ohpen; more than any of it. Getting everyone on the same page. It was something not a lot of people could have pulled off," Chris recalls.

While proud of the achievement, Chris credits all the parties working together to make it happen. "Amazon, Robeco, and Lydia all worked with me to make it happen," Chris says. Chris is also extremely proud of Lydia. "Can you imagine, working all night just before Christmas? All the offices were dark, but there was this little light in the corner and a young twenty-seven-year-old sitting there all night, working alone. Meanwhile, she was taking shit about it the whole time because her friends were like, 'What the fuck are you doing? Working all the time?'" Chris admits she was a huge asset. "She managed the paper trail. I might have said, 'I think I saw that before.' And she would say, 'Oh, let me check. Yeah, version thirty on the fifth of August—we all agreed.' And I would say, 'Okay, great, great.' It really showed me that she could step up to the plate. And it catapulted her. Some people take small, incremental steps, but this really pushed her forward quickly. I know she will remember this project until she dies. I'm a hundred percent sure that when she's eighty and she has her grandkids, she will think of this. Because this was not only a defining moment for the company but also a defining moment for all the people involved."

When Chris finally arrived in Switzerland in the morning, he exhaled. "It's done. Whew, we did it." Chris also knew from that moment on that Ohpen was here to stay. Everything that happened afterward was going to be a piece of cake compared to this. Chris looked over at his wife, and she was still asleep. "I can still get emotional just thinking about it. It was the absolute tipping point. That was the moment I knew we would make it. It was also a moment of pride. I was very, very proud that I got it done. It was one of the proudest moments in my life. But again, the funny thing was, in that moment there was nobody to share it with. I shared it with myself. I was in the car and thought, '*Whew, dude that's something. We got it. We'll be here forever.*"

And Lydia was by herself too. In the cold, dark morning, she hopped back on her bike and rode home all alone. And Lydia, like Chris, was happy and proud too. She could feel Chris's approval and pride as well. "I felt he was really happy that he could rely on my being in the office and trusting my review of the documents. That's something I remember. And I think it was like a tipping point for both of us. We pulled it off together."

To this day, both look back fondly on the memories of that night. In fact, Chris goes so far as to say, "My biggest achievement in my twenty-year career is not Ohpen—it's that we got this contract approved." After nine months of tense negotiation, AWS, Robeco, Rabobank, and the DNB approved Ohpen to be the first service provider using cloud technology—the first of its kind in the world.

CHAPTER 7

ONE THOUSAND REJECTIONS

DURING THOSE FIRST BUSY TWO YEARS, while the team was back in the office building the platform, Chris was also flying around the world to pitch the first cloud-native core banking engine in the world to banks. During this whirlwind time, Chris, sometimes joined by Michel, met with what felt like *one thousand* bankers, though in truth he knew it was more in the hundreds. Whether it was five hundred or a thousand was no matter, though, because he was rejected by all of them. Some rejected him because of the oft-repeated refrain, "The cloud's not safe," and bank executives refused to believe otherwise. Some loved the idea but didn't want to risk being Ohpen's first client. "Call me when you have five clients" was something Chris heard a lot in those days. Chris already knew that if he was going to sell the Ohpen platform, he needed a proof of concept, and his team was busy building one. They didn't just build a website, though— they built a full-scale asset manager. As just a stand-alone proof of concept, ohpen.nl managed to service customers who invested €125 million worth of assets. In order to do this, they created several index investment funds. And they also built a tool—a robo-investor— that would propose options to customers on how they could invest their money. This robo-advisor, the first of its kind, was later sold to the clients of Ohpen. The asset allocation process depended on certain criteria that customers would fill out, and then the tool would automatically invest the money in the appropriate index funds and rebalance it, daily if necessary.

For a start-up fintech company with less than a dozen workers, building a workable asset manager was one thing, but going for a full-on company takeover was another. "We were running Ohpen. We were meeting with bank after bank to try to sell the platform. And all the while we were building this platform *and it was working,* but banks wouldn't budge," Chris recalls.

Needless to say, Chris was busy. In the process of building his own company in order to prove its worth, he essentially had to build an asset company *and* acquire an insurance company—not to mention streamline their processes and cut costs—all just to convince banks to work with Ohpen.

Chris's job then—trying to get clients—was far from easy. He didn't just have to convince a bank's board; he had to convince all the stakeholders that it was in all of their interests to switch to Ohpen's platform. In order to get all the stakeholders on board, they had to help them make a paradigm shift. Again, not the easiest thing to do. On the one hand, they had to deal with people who didn't even know what the cloud was—sometimes Chris had to explain things so many times that in the end they almost gave up—and on the other hand, he had to deal with "experts" who thought their way was the best way. It was exhausting.

Michel remembers, "Chris and I spent many, many hours, weekends, and evenings traveling together. I joined him in a lot of these meetings in London, in Switzerland, and … well, *everywhere.*" Michel was mostly there for moral support or as a sounding board. It was Chris who was in charge of selling, and Michel admits it wasn't always fun to be there and witness it firsthand. "The sales process within a bank is gruesome. It's always uncertain what's happening inside the bank, because we were dealing with ten different departments, or twenty different people, and they all had an impact on the process. And Chris had to manage them all. Chris was basically doing it alone," Michel adds. "That's why I think it was against all odds, because normally this would not be doable by a small firm. I don't even think it would have been doable by a big firm."

Michel, having worked in banks himself, knew exactly what Chris was up against—internal politics, personal agendas, and an overwhelming resistance to change. Michel also knew that Chris wasn't going to be every banker's cup of tea. "He was young and sometimes a bit assertive but always very intelligent, always very knowledgeable," Michel admits. "But he knew what he was talking about. Some people really didn't like the cockiness, but others thought, 'Hey, this guy knows what he's talking about.' So it was love him or hate him, and we noticed it." Michel never tried to rein Chris in. "That would be crazy, because the guy has a talent. You have to give the talent space to grow and do its thing."

In the end, it became clear, though, that Chris needed a counterweight in his sales approach, something to balance out his extroverted nature—he needed his old friend Matthijs, who was perceived as more of an introvert. Chris once again asked Matthijs to join the Ohpen team officially. Two years into the new venture, Chris knew that one person was missing to make the team complete. If he was going to seal the deal and lead a team and take the product to the market, he needed his partner, Matthijs, to help him. He gave his friend one more shot. Chris knew if he was going to take Ohpen to the next level, he needed Matthijs's skills and expertise. This time, Matthijs said yes. After a lot of meetings with Chris about the role, not to mention with the major investor, Michel, he was officially approved by the board of directors, and he officially resigned from BinckBank on his birthday—August 29, 2011. Matthijs had left his beautiful apartment in his beloved Paris, his weekend home in Normandy, and his office view of the Sacré-Coeur, not to mention the safety of the "golden cage"—job security for life—and embarked on the greatest risk of his career. He and his wife, Siena, packed up their homes, their one-year-old daughter, Sophie, and their newborn son, Gijs, and moved back to Amsterdam.

There was no time to spare. The two men had to get to work immediately. Chris started to make profiles of each bank and divided the sales pitches based on personality and "the right fit." They would decide it they needed Matthijs or Chris or both, and then they would determine who would be lead guy. "Ego was never a part of these decisions; we just wanted to sign the client," Chris adds.

Michel eventually backed off attending sales meetings. Chris and Matthijs were more than capable of handling it. Two years into building the platform and trying to sell the platform, the pressure was on. Chris was living out of hotels and in airport lobbies, running the office from the road. But he was unwavering and didn't give up. He just needed that elusive manta ray in the giant ocean of possibility: forward-thinking, agile, open-minded, money/savings-conscious, future-minded people who were unafraid of risk and wanted to be on the vanguard of change. Chris believed there were bankers out there who just "got it." All they needed was one who was hell bent on "moving." As Michel says, "In banking there's no way you cannot move. If you're not moving, you will not survive. I think that's a real must for banks. I'm not sure all the boards understand that their survival is at play today, because they're used to making money so easily, and they never had to be real entrepreneurs. They had to just manage what they had, and that's how they made money." But, Michel acknowledges, change, whether boards are ready or not, is coming. "You now see entrepreneurs like Chris coming in, changing the whole environment." Prior to Chris and Ohpen, many banks thought they were innovating when they were building apps to help with the client experience. But, Michel adds, banks essentially "didn't know what they didn't know—they had known and unknown unknowns." And he adds, "The back end of these banks is a monstrous machine that they know is going to fail at some point in the future. And if they are not really working on changing that fast enough, it will fail in a variety of ways. In some cases, it has already failed for some from a security perspective."

The ironic thing about shifting bankers' paradigms was that while they were telling Chris and Matthijs how *unsafe* the cloud was, their own systems were already highly compromised. Michel adds, "The whole fact of the matter is that the systems the banks use are, in many cases, outdated from a modern-day perspective; the banks look at the risks of new ideas, or they didn't have that in their minds at all."

Convincing banks that their current systems were outdated and that Ohpen was agile, fast, safe, compliant, reliable—and could save them a lot of money—was the hardest thing Chris and Matthijs had to do. From their perspective, it should have been easy. No one would have argued with them if they'd said modern cars, which are equipped with better tires, faster engines, and safety features, are better than cars built in the sixties. No one would have argued with them if they said that iPhones were better than the first clunky, unwieldy, and unreliable mobile phones with spotty service of twenty years ago, and yet when it came to their own legacy systems, they did.

Building the platform was a success. Building the company, even acquiring and running another company as well, had been a success. But getting clients was proving to be nearly impossible. Before signing Robeco, Chris almost lost heart. As the two-year mark clicked by in early 2011, Chris was well aware that the money was running out. Looking back now on that time, Chris can see how close he was to all of his work being for naught. But he never thought for one second of quitting or throwing in the towel. "It's just that I don't quit. I don't give up. I always said that if I can survive an absent, alcoholic father, a youth with a lot of sorrow, or my own drug addiction as a teenager, then all of this was nothing; it was an easy afternoon for me." Though the constant rejection was disheartening and what Chris calls "tough," his mentality to keep things in perspective kept him going—well, that and two people, one of whom was his wife, who always had his back during those long, dark, and lonely days. "So many times I was in London or wherever, and all I wanted to do was fly home—and something happened and the plane would be held up—and she would just be there. All these years she felt my stress and always had my back." The other person whom Chris relied on heavily was Michel—who repeated to Chris often, "It just takes one." Chris believed that as well. All he needed was one—just one bank to believe in him; just one bank to take a chance. "You just need one," Chris adds emphatically. "You just need one person to have your back; you just need one client; you just need one idea. You just need one good worker. You need to start with one."

WHY IT IS SO HARD TO SELL: STAKEHOLDERS

Just why was it so difficult to convince banks to change to a safer, more reliable, and compliant core banking engine that could save banks and their customers millions of euros? It had a lot to do with the stakeholders. In the most simplistic terms, there are two types of people running banks. First, there are bankers that are just "stewards," or what Chris calls "babysitters." They're not necessarily looking out for themselves, but they're not looking out for the bank's long-term interests either by being extremely risk averse and conservative. Then there are those rare "mavericks"—bankers who are looking far out into the future; they want to save their customers' money, and they want to make sure the bank lasts long after they are gone by innovating. They are thinking beyond their own short-term interests and also don't want to be responsible for being the Kodak or the Blockbuster of their industry. Technology changes the world at a very fast pace, and if they don't adapt to those changes, they know they will not be there in the future—*and, of course, they don't want to be responsible for this.* Within these two categories are a variety of stakeholders, however, who each have their own specific reasons why they don't want to outsource or change the way they do business as usual. Additionally, the world was still in the middle of the financial crisis at the time, and some banks just had other priorities.

Rather than be dissuaded by all the potential challenges with having various stakeholders involved, Matthijs sees it as a bit of a game. "I look at closing a deal like a game of chess. Not in the way that the prospective client is our adversary but in that we need to know what all his 'pieces' do and when to make the right move."

Similar to any complicated game with a worthy opponent, the "game" can take quite some time. Matthijs knows that the key is patience and dealing with the full scope of the deal. "Like any game or competition, it can be intellectually challenging, but contrary to sports, brute strength won't bring you anything. A small piece can take down a more important piece, and as in all games, each player brings with them various skills, expertise, tasks, agendas, and even blind spots—and one has to be able to handle them all."

The first area Matthijs considers is the legal environment. "We are talking about regulated environments. For certain operations, you need to either inform the regulator, or you may even need to obtain their approval. The latter is called a declaration of no objection. The funny thing is that regulators do not say it is a good idea, but they won't object. It illustrates the position of the regulator and that they hardly ever or never say what they really think. The best you can ever get is that they don't object," Matthijs says. The consequence of the system of permissions and licenses is that most financial service providers consist of more than one legal entity that together create a group. The holding company that is on top does not necessarily have its own permissions, but the working entities always have them.

Each entity (see the chart below) has its own board that holds responsibility vis-à-vis the regulator. They can be held accountable from a personal liability perspective, so they usually take their jobs seriously. "Besides the liability part," Matthijs adds, "if the regulator doesn't like you, they will refrain from 'not objecting' so that you can stay a statutory board member or that you get a comparable position elsewhere."

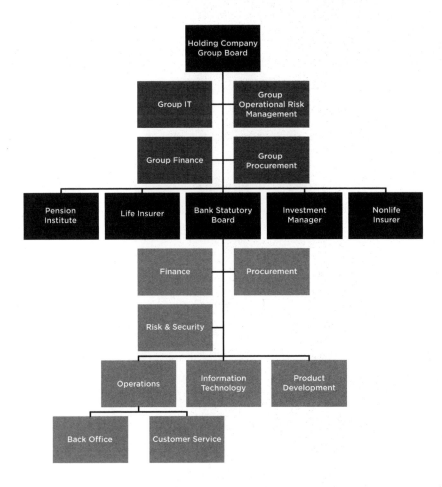

The board would typically consist of one or more of the positions as outlined below.

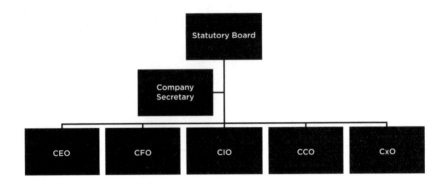

When closing deals, Ohpen works with the various stakeholders in the form of *organizations*—the company itself that they want to sell to, the group in which the company operates, the investor(s) in the company or the group, the accountant and auditors of this company, the advisors of the company, the supervisory board of the company and the group, and the authorities/regulators (which include the prudential regulator, conduct regulator, privacy authority, and fiscal authority). "In the end, the statutory board members of the company itself—most of the time at least two of them—are the single most important for us," Matthijs adds. "They will sign the agreement to outsource their account administration to Ohpen. In most cases, nobody else but them can sign. But they will only sign after they have gone through all the necessary steps to do so."

And these steps often include convincing all the other various stakeholders involved, including the board. The CEO of the company is usually the leader of the pack. He is the one in the media, the one talking to investors, regulators, and so on. "Since the CEO is usually not the one who handles the outsourcing itself, he or she is not our 'go-to person' within the company," Matthijs says. The CFO is the one under whose supervision the business case is made to prove that outsourcing will also benefit the company. The business case should obviously be favorable for Ohpen. It is therefore important that the right information is provided to the CFO. "This can be a struggle. Not all relevant costs that make up the total cost of ownership are taken into account," Matthijs argues. "Sometimes the team under the CFO compares apples to oranges. Ohpen needs to make sure the business case is sound." In the majority of the deals, the COO has been the most important stakeholder for Ohpen. This person is responsible for operations, and that also includes IT in a lot of cases. In the end, the departments that profit or suffer from the implementation report to the COO. In fact, Matthijs credits the COOs—Leni Boeren from Robeco, Sjaak de Graaf from Nationale-Nederlanden, and Alexander Baas from de Volksbank—as having Ohpen's back during the decision-making process as first movers in cloud banking.

Just because Ohpen could have a great working relationship with a COO, they can't overlook other key stakeholders or supporting departments. Risk management is the most important nonoperational department for Ohpen. Not only are they involved in creating the ENISA risk self-assessment; they also write a more general report on Ohpen. This report gives comfort to the board that everything is properly arranged.

In order to bring the board comfort, security can't be overlooked. All the security officers tend to think that their security measures are the best in the sector. They want to impose the security requirements of the financial institution they work for on all their different vendors and suppliers. They also need to sign off on the deal. Building a relationship with the security team is crucial. And equally crucial is the relationship with the legal department. The person (or in some cases persons) responsible for legal and compliance will want to make sure that the scope of work meets the real requirements and that the correct clauses are in the contract.

No department is too small or should be underestimated. Procurement, for example, takes care of the selection of vendors. Most of the time they create a policy that all vendors need to subscribe to. The procurement manager usually handles the process and makes recommendations but is not the one who decides. Nevertheless, Matthijs and the team work with them. No internal person's role can be underestimated.

The great irony of all this, however, is that the departments that *actually need* to work with the supplier are not attributed a leading role in the selection process. This doesn't go unnoticed by Matthijs. "They are involved, but I have never really seen that they have the last say or that they are involved from the start and determine the course of the project."

The reality is that the back office needs to work with the Ohpen platform. The IT people need to connect to the platform and interface with it and integrate it into the rest of the architecture of the bank. The front office (contact center) people also work with the platform on a day-to-day basis.

Another group that the team considers when working on a deal is the group in which the company operates. Matthijs adds, "In many cases, our clients are part of a group, a conglomerate, if you will. Financial services products in different entities are combined in a group—for example, bancassurance." The group sometimes is under direct supervision or is the entity that is listed on the stock exchange. Most of the time, the group has no operational activities of their own, but they execute the overall strategy and handle the consolidated finances, including M&A (mergers and acquisitions) and ECM (equity capital markets). They carry a lot of responsibility and therefore have their own supporting departments, like risk and security. "What we have gone through in the past is that usually the board of the group was not directly involved, nor did they sign, but the group departments for operational risk management, security, and procurement were involved, and all needed to sign off on the deal," Matthijs says. Signing off on the deal means not that they sign the agreement itself; they sign an internal piece of paper saying that they have verified that the new supplier complies with all the group policies.

Investors are almost never directly involved, but they are important stakeholders because the board of the group invariably needs to report to the investors and prove that they are making the right decisions. The board is accountable and reports to the general assembly of shareholders. The board is discharged by the shareholders, so they keep their interests at heart at all times.

Accountants are hardly ever involved presigning. But they are stakeholders because they sign off on the annual report of the bank. And the bank relies on the data obtained from our platform to create the balance sheet. The accountants themselves must verify if the underlying administration of the bank is correct, unless they get a copy of the SOC1/ISAE 3402 report from Ohpen's auditors that the Ohpen platform functions correctly and that they can rely on its contents.

Banks are not afraid to obtain advice and often seek out third-party advisors. Working with either the "big four" (PwC, EY, KPMG, and Deloitte) or the likes of McKinsey and Co, Boston Consulting Group, Bain, or Accenture not only provides new insights and benchmark information but also contributes to the risk management of the board. From Matthijs's perspective, seeking advice from outside is some kind of "get out of jail free" card. He says, "If the board had obtained advice that outsourcing to company A was a good idea, and when it turns out the other way, at least they can say that they obtained advice elsewhere." Because of this, Matthijs says Ohpen spends a lot of time making sure that the leading advisory firms are aware of what Ohpen can do. "When we come across an advisor within a bank, we then already know them," Matthijs says. Secondly, sometimes it is not a bad idea to ask an independent expert what he or she thinks. It is just about making this huge decision in the best possible way.

For regulated entities, which are always so-called public-interest entities, a supervisory board is mandatory. The supervisory board meets with the board at least four to six times a year, and often all kinds of subcommittees are created, like the audit and risk committee or the remuneration committee. In some circumstances, the executive board needs to get approval from the supervisory board before they engage in certain strategic projects, like outsourcing. Sometimes Ohpen actively engages with supervisory board members, but most of the time these processes are behind the curtains for them. Even though they don't get a lot of face time with these entities," Matthijs says, "the supervisory board members are undeniably an important stakeholder for us."

Among the most important stakeholders are the authorities. "Banks really pander to the authorities, no matter which one," Matthijs says, being someone who knows and has worked in a bank. "When addressed by the authorities, bank officers say, 'Yes, sir,' 'No, sir,' and answer the phone or reply to emails directly. You don't want to mess up the relation with the supervisory authority," Matthijs says.

The prudential regulator (PRU) determines the capital requirements for the banks. Although no direct impact is made by outsourcing, banks must maintain capital for certain risks, of which outsourcing can be one. The PRU is in some cases like the Netherlands (in the UK the conduct authority handles this) responsible for enforcing the financial law, including its articles on outsourcing. This makes the PRU a very important stakeholder. It has the power to forbid banks to work with certain providers. It creates risk assessments and can act to make sure banks take certain precautions. Banks need to inform the PRU if they want to outsource. The banks need to make sure that the PRU can still exercise its supervision and that outsourcing is not an impediment to this. "The DNB commented on concentration risk in cloud providers recently. Of all regulations, this one is the most important for both the banks and for us," Matthijs says.

The conduct authority makes sure the interests of the consumers are guarded by the banks and that the banks "behave" properly. It acts if it feels the products the banks create or the services they provide to consumers are not in the interests of the latter. In the Netherlands, it is the Authority of Financial Markets (AFM) that is involved in the interpretation of European laws like Markets in Financial Instruments (MiFID) and in that way imposes certain requirements on the Ohpen platform, like the disclosure of fees paid by consumers for investment services or investment products.

Privacy is becoming more and more important, as well. The General Data Protection Regulation imposes a new set of rules on banks. Even more than before, they need to be able to demonstrate that they guard and protect the (personal) data of their customers to the best of their ability. The Authority of Personal Data (AP) is not involved in precontractual obligations, but the lawyers of the bank make sure the agreement complies with the law and that the personal data is properly protected in the servicing.

One of the stakeholders that is more important for Ohpen than for the banks, according to Matthijs, are the fiscal authorities. The ministry of finance determines the VAT law and income tax law, and the fiscal rules determine a material part of the development of the Ohpen platform. If either the parliament or the fiscal authorities make changes to the laws or technical specifications of reporting, this means Ohpen has to adjust its platform—hence a lot of work for the team.

GETTING THE KEY STAKEHOLDERS ONBOARD

Of course, the problem inherent in having so many stakeholders is that everyone ultimately has their own agenda. Often the most resistant to outsourcing is not surprisingly the internal IT department. Sure, they are well aware of the faulty processes and patches in legacy systems. However, many worry, quite naturally, whether or not their jobs will be eliminated if the core banking engine is outsourced to a company like Ohpen. "One time, an employee of one of our clients said that he had to fight this, because if the bank went for it, then he would not have a job anymore," Erik recalls. Chris adds, "When banks host their own software, going with a SaaS company means they don't have to host themselves anymore. They can do the math."

Chris had people come up to him and admit, "If I agree to this, I will not have a job anymore." Others admitted reasons relating to personal and internal politics. They said to him, "My boss tried to build this and failed, and we had to write off thirty million euros. If the higher-ups in the US hear about you, then *he* will lose his job, and so, in sum, my boss will never sign this." They're also hesitant for unselfish reasons too. They are well versed in compliance and risk and are worried about preventing data breaches, unwanted access to systems, viruses, and other threats. If something happens to the system, some IT employees naturally worry about where the ax will fall when the shit hits the fan. Risk and compliance officers feel the same way. Their main job is to ensure that any outsourced program aligns with all banking policies, procedures, laws, rules, and regulations. At the same time, development operations worry about the need to change all their procedures and processes. Operations people aren't always wired to naturally rethink the processes and try to map a paper-based world on a digital world; they're used to things being a "certain way." This makes it a bit difficult for Ohpen, because a main driver within Ohpen is to *rethink how to do things*. Chris recognizes the difficulty in translating Ohpen's vision and way of doing things to people who feel safer maintaining the status quo. "Many times it comes across as we are saying *they* are doing it wrong," Chris adds. "So we have to be wary of those politics and emotions when we show our software." He recalls that one of the most frustrating conversations ever was with the head of operations of a very large bank. He said to Chris, "You have it so easy with everything automated! I have to do everything manually." Chris was in disbelief. The head of operations didn't like Ohpen *because his platform actually worked and was automated!* "My platform can do with eight people what you do with eighty," Chris argued. "Of all the reasons to say no! The one reason you chose was *I don't like it because it works?*" Chris was so frustrated he almost gave up that day.

Chris believes there are two types of consultants/advisors: those who have their clients' interests in mind and those who don't. It is that simple. Chris remembers meeting with one bank who paid their consultants based on the number of incidents they solved! The more a legacy system failed, the more money the consultants would make. However, Chris

> Chris believes there are two types of consultants/ advisors: those who have their clients' interests in mind and those who don't. It is that simple.

believes, there are plenty out there trying to help their clients save money and be better.

All of these stakeholders make their recommendations to their board of directors. If the board of directors gets bad advice from any of the above-mentioned stakeholders, it could end the conversation before it starts. Again, these stakeholders aren't necessarily looking at the outsourcing option objectively, because instead they are looking out for their own department's interest. If that is the case, decision makers will be hesitant to migrate as well. Even if it makes more financial sense to outsource, that doesn't always guarantee decision makers will recommend outsourcing.

This is why making a business case is so critical, as is having objective IT people and ops people assist in making this case. Chris has reviewed many business cases that were blatantly wrong, mainly because they did not include actual costs. Chris recalls one bank claiming they didn't have to pay for "hosting." Chris pointed out they *owned* a data center where they hosted their software. *Of course they were paying for hosting.* "Yes, but it's not my P&L," Chris says mockingly. Another time, Chris approached an asset manager and said he could save him €6 million by switching to Ohpen. Yet in the business case, the asset manager pushed the hosting and operational cost as an expense for their retail customers. The asset manager put the cost of IT and operations into the mutual fund itself. This means that their clients were paying for this. Chris explains, "Let's say a fund has fifteen billion euros in assets, and the total cost of this fund is one percent. That one percent gets taken from the fifteen billion every year in order to cover the cost of the mutual fund. But what this asset manager did was put all the cost for their retail operation into this fund as well. So on top of the 150-million-euro cost for the fund, they would add the eight-million cost. *No one sees this.* This asset manager actually said that his retail operation was for *free*. If you look at it this way, our business case was wrong. It was not an eight-million cost versus a two-million payment to Ohpen. It was zero million for him versus two million for Ohpen. He was such an idiot. The laws should be changed so that *nowhere* in the world can these costs be made to be paid by a retail customer."

Despite all they were up against, Chris and Matthijs believed wholeheartedly in the Power of One—one platform, one vision—and in themselves.

PART FOUR

IN THE DEEP

(2013–2014)

CHAPTER 8

MIGRATING THE FIRST CLIENT

AS SOON AS THE TEAM DELIVERED the proof of concept to Robeco for their approval, Chris wanted to reward the entire team for all their hard work on getting the company up and running and finishing the Robeco proof-of-concept phase. He went to Michel and made his case to spend €50,000 on his team in such an extreme way. "These guys have given their all. For two straight years they worked day and night, seven days a week, to build the platform—and they still have a long way to go—some twenty months to go live and fully migrate them," Chris says. Even though Chris was "in the hole," he made the case for rewarding the employees and showing them his gratitude, and Michel didn't hesitate to agree. "I wanted to do something that the people couldn't afford by themselves," Chris adds. And that he did. He spent €30,000 on a massive and state-of-the-art ski chalet, then another €20,000 on flight tickets, ski lift rentals, ski lessons, and an on-site chef and housekeeping team for fifteen people. Chris wanted to do something nobody else would ever imagine doing or could do—at least on their own. He was also being pragmatic. He acknowledges, "We weren't paying overtime or offering a pension or benefits—nothing. So I calculated what it would have cost if I'd had to pay for eighty hours a week instead of forty, with pensions, benefits, and insurance. We would not have made it. We would have run out of money. And I think it is a two-way street. Employers and employees both have to put some energy into it. When there is a balance, sometimes great things can happen, and that is so abstract, you can't really explain it. It is just energy—positive energy."

For five full days, the team reveled in every new experience—having a wait staff, for example, was a revelation for many of them. "Dude, what the fuck is this? This is great!" is something Chris would often hear throughout the week. "Who has a staff and a cook at twentysomething years old?" Chris asked through a laugh, enjoying the memory of watching his young team experience so many new things at once.

On the first day they all went skiing together and had a blast, but then a massive storm came in, and they spent the remainder of the week indoors, bonding and getting to know each other outside of work. They ate every meal together, played cards, watched movies, talked about the organization they were building and the technology of the future, laughed, and in many ways acted every bit like brothers (and one sister—Lydia was the lone female). In many ways, Chris created a family and a home like he had always missed as a child. He loved these people and wanted to show them gratitude. In the evening, as they sat around a massive table, Chris took the scene in. He was extremely grateful for them. And it was mutual too. Lydia, Erik, Bas, Ilco, Andre, Jan-Willem, and others can still vividly recall the intense feelings of joy, camaraderie, and appreciation from that first trip. "I will never forget that trip. Those are amazing memories," Lydia says, recalling it. Erik adds, "The cool thing for me was that it actually felt I was on holiday with friends, not colleagues."

The trip impacted the group exactly the way Chris wanted it to. The team bonded and felt in many ways more dedicated than ever to give it their all. When the trip was over, they returned to the office and resumed working around the clock for the next twenty months to get the Ohpen platform ready to go live for Robeco. Back in the office, there weren't a lot of pats on the back or time to celebrate. It was business as usual, and no one complained. Everyone felt like they were very much in it together.

ON THE MOVE ONCE AGAIN—MOVING
INTO THE ROKIN OFFICE

The months between the end of 2011 and the end of 2012 were tumultuous, and the company was experiencing a lot of changes and growth. In addition to working on getting Robeco ready for migration, Ohpen was growing and needed a larger space to operate in. Optimistic about their future and growth potential, they agreed to expand. Matthijs saw that the building Rokin 111 was available for rent. It is a majestic four-story building from the art deco era, built out of massive stones and decorated with custom iron grillwork and detailing, including an etching of a large barrel at the top. (The building originally served as a gentlemen's club, and it was a spot for local professionals to gather, meet, and of course drink.) The tall windows on each floor overlook the busy Rokin neighborhood, which is in the heart of Amsterdam. Standing in Chris and Matthijs's office, one can see the train station below, the trams that run down the street, and the historic stepped-gable buildings that surround it. It is quite literally smack dab in the middle of everything that is happening. It represented everything Ohpen was—very much part of the past and the present of Amsterdam—and, more importantly, the future. Matthijs and Chris, who wanted the building to be an "experience" and to be "different," immediately fell in love with it. They knew any potential client who walked through the doors would feel the same way.

At the time, Amsterdam was still reeling from the financial crisis, and many office spaces remained vacant. Chris told the real estate agent in charge of the search, "I don't know anything about real estate, but if it's anything like the stock market, prices go up and down." He used the financial crisis to his advantage. Knowing the space was going to be sitting empty, Chris approached the landlord, who he knew would be looking to cut a deal to get someone in the space. Chris recalls saying, "You can go look for someone who will pay more rent, but if you're like me, if you're more interested in having an income every month and not going for max profit, then you should go with us. You will love us as a tenant, because we always pay our rent, we will invest in the building, and we'll always make sure it's spick and span every second of the day." Chris told the real estate agent what he was willing to pay: "Take it or leave it." They took it.

FINISHING PHASE ONE OF
THE OHPEN PLATFORM

While Chris and Lydia were enjoying the successful contract negotiation, Erik, Bas, Ilco, and their old BinckBank colleagues—Andre Janmaat, Anton Zhelyazkov, and Raymond Morsman—were working for twenty months nonstop, seven days a week, sixteen hours a day to finish the platform. Though the team was small, it was an optimal group, and everyone was at their best—driven, brilliant, and 100 percent committed. Even the interns. One who demonstrated the fierce fighting spirit was Robert Siertsema, who joined the company in February of 2012. He participated actively in the gym classes (which were provided by Ohpen in the basement gym of their new building on Rokin). During one kickboxing session at the end of his internship, he sparred against colleagues that were better than he was. But he kept on going. At the end of the session, his tongue still hanging out of his mouth, Chris offered him a job, because Chris was convinced that someone who trains the way Robert did would make a great colleague, embodying the *give it all* mentality.

They needed that fighting spirit, because between April 2012 and 2013, the team worked around the clock to get the platform up and running for Robeco. "Every month we had to build, and every month we had to demo," Erik recalls of the period working with Robeco. At the time they had a large auditorium in their office, and every month Erik had to open the platform and demo the new features the Ohpen team were working on to the Robeco team. There would be anywhere from 180 to 200 people in the room. "It was so stressful," Erik remembers. "Every month we finished just two or three hours before I had to demo." Erik recalls Ilco and Bas advising him not to touch certain features or giving him the go-ahead to demo while he was in the car on his way. "It was like having a heart attack every time I logged in. I died twelve times, at least, that year," Erik exaggerates but shakes his head, recalling all the stress. In the end, the team pulled it off. They may have worked around the clock and down to the wire every time, but in the end they built a platform that worked and that was ready to launch.

By April 2013, Ohpen and Robeco were ready to implement the Ohpen platform. This entailed Ohpen having to migrate a few hundred thousand accounts and billions of euros in assets from Robeco's old platform to Ohpen's newly built one. Now it was Erik who was feeling the heat. "If we'd made a mistake somewhere in building the code over the past twenty months, we had no one to fall back on. I was the one checking everything, because I was the only one then with all the knowledge to be able to check it. So I had no sparring partner. I knew the client had tested it as well and that it should work, but I thought to myself, *If I made a decimal error, then instead of eighteen billion, we would have eighteen million.* I didn't sleep for like two weeks leading up to the migration."

Erik also put on twenty kilos—he was living on takeout and sitting in the office for seven days a week and not exercising. Looking back on that time, Erik has nothing but happy memories, though. "I really enjoyed it," he says with a laugh. It was, Erik recalled, a harmonious time. "We were a small team with one goal." And that ultimate goal was to migrate Robeco over the Easter weekend in 2013.

All hands were on deck that weekend. "We had the run-book in front of us. We started at nine on Saturday. The entire day was broken down into five-minute intervals." This precision and attention to detail was required. Nothing could be overlooked. Fortunately for everyone, on the team was another man named Joost van de Ouderaa, who was so detail oriented, Ilco joked, that "if he kissed his wife, he'd check it off a list and move on to the next thing: 'I gave her a hug. Check. Done. Next.' But that's the kind of guy you need on a job like this."

For the most part, there was little risk involved that the migrations wouldn't be successful. Though it was a tense and pivotal time, Ilco was so confident in the platform that he didn't sweat it. They had run so many tests and run-throughs that the actual migration was a formality. Ilco also knew it wouldn't take the entire weekend the way previous migrations had taken at BinckBank on the legacy system. "The actual technical migration to get all of Robeco's data only took an hour." What took the entire weekend was ensuring the security and doing system checks. There were bound to be hiccups—calculation errors, incorrect values, incorrect performance data from the customer, and other standard issues when migrating large amounts of data. The team was there mainly in case catastrophe struck, which never happened. However, at one point there was a glitch, and Andre was the only one who knew what to do. It was three in the morning, and he was nowhere to be found. The team raced around the office looking for him. Lydia eventually found him sneaking a nap under Chris and Matthijs's boardroom table on the fourth floor. Andre woke up, raced down to the room, reviewed the error, and quickly fixed it. Andre wasn't the only one who fell asleep, though. Some team members dozed off on chairs; others slept in the gym or under other tables or desks. All in all, the migration was largely uneventful. Most people who were there at the time can't recall a specific incident that felt catastrophic. There were no "Oh no, this will never work" sort of moments at all. Everyone was there mostly to fulfill their testing roles and to make sure everything ran smoothly. So Erik and Ilco didn't mind too much if team members napped if they weren't needed.

However, Erik didn't sleep at all. "Once we migrated everything, we ran a copy environment. I ran all the core processes, and I personally checked all the output on the copy environment before I would give my 'all's good.'"

Erik was vigilant, checking everything not once or twice but three times. The Robeco migration became a template of sorts for all future migrations, though now everything is automated and it's a well-oiled machine. They built within the migration platform every foreseeable issue as well as workarounds to deal with them. They also knew how to approach escalations and what protocols to follow should they need them (they didn't). Kees Postma, who was an intern at the time and working during the weekend of the migrations, recalls, "The migration went flawlessly." He was impressed by the competence and outward calm of everyone—especially the leaders: Matthijs, Chris, Erik, Ilco, and Bas. Though Erik and others may have felt the stress internally, they never showed it or expressed it outwardly. The entire weekend was a lesson in *how things should be done*. They planned for every scenario. They practiced it and had run-throughs, and then they executed with precision according to the detailed run-book. It became the working template for all future migrations, and it proved that with the right infrastructure, great software that just works all the time, and an excellent and well-trained team, a migration could be executed without incident.

By Sunday morning, Erik and all the team members who were there were rendered speechless when Angelique and members of her team who weren't working on the migration surprised the guys with a full Easter brunch fete. Angelique insisted that no one "buy anything," that everything should be homemade. Her main objective was to send a message of gratitude to the hardworking migration team. She recalls telling her team, "We're going to prep everything to show them we know what they're doing, we know what they're working on, and we are here for them, because we're one." They each made select dishes and put on what she deemed an "extravagant" brunch for all the guys who'd spent the holiday away from their families. It was a moment of intense solidarity and celebration, albeit short lived because there was still work to do. Robeco needed to go live, and so that, too, took a day of testing and oversight. By six in the evening on Sunday night, Robeco's team told the Ohpen team, "Everything looks good. We're ready to go."

Much the way Chris felt when he got the call from Robeco that he had won the RFP process, Erik was relieved when he heard the news but too exhausted to react. The entire team felt the same. There was no party, no fanfare. They congratulated each other briefly, and then everyone went home to get a good night's rest. They all had to be back in the office the next day. Erik still had one more meeting. He had to speak to the Robeco board of directors at midnight to verify that the job was complete and to assure them he would be in their offices the next morning.

The following morning, Erik still remembers when he arrived in the Robeco offices. Tired, he entered the elevator of the bank's high-rise building and took the long ride up. Then the doors opened to a call center. Erik stepped out of the elevator and could hardly believe his eyes. He wasn't tired anymore. A surge of energy shot through him. He felt goose bumps all over. Before him was the Rotterdam skyline etched just outside the large bank of windows that spanned the enormous room. And in front of the beautiful skyline was something altogether more beautiful to behold. He saw a sea of desks, each of them with computer monitors displaying the loading screen of the Ohpen platform. His life's work, his team's hard work, Chris's vision—all of it—was there staring right back at him. *They'd done it.*

However, Erik knew their work wasn't finished. "We all knew that from a functional point of view, that product wasn't ready to serve a lot of clients or serve larger savings banks." In other words, as soon as the team had a chance to celebrate their success, they were already thinking about the next level and making improvements and expanding.

Chris, however, wanted to show his gratitude to the team that had worked around the clock. As a reward, Chris rented another chalet and took the migration team on a ski trip. For Chris it was vital that he show his gratitude and appreciation. There was no end to what the team members would do for each other—within or outside of work. At night, when sitting around the large table and enjoying views of the mountainside, Chris was able to sit back and relish what he had accomplished. He hadn't just built a platform; he had built a family. He loved each one of them and was so inspired and infinitely grateful for all they had done. He also felt a point of pride. Some of the team members had grown up right before his eyes and had accomplished more than they themselves had believed possible.

CHAPTER 9
THE SHARKS CIRCLE

WITH ROBECO SIGNED and up and running, Chris could finally come up for air. For the better part of three years, he had been on the chase—looking for funding, looking for stellar partners and employees, looking for clients, and even looking for companies to acquire. But by the middle of 2013, Chris found himself in entirely different waters—it was Ohpen that was being chased. Word had spread: Ohpen was a serious contender in the sea of competing fintech companies. As Chris, Matthijs, and the team rose to the surface to enjoy their success, they noticed they weren't alone. There were sharks in the water, circling nearby and seeing if they could take a bite out of Ohpen's success. They weren't just any sharks, though—these were Great Whites, the top of the proverbial fintech food chain. Apparently, Chris's idea had turned the heads of some executives at one of the world's largest banking software companies. At least three different large core banking companies approached Chris with the intention of buying Ohpen. One company started the negotiations rather quickly. This company served over three thousand clients, forty-one of which were the top fifty banks in the world. All told, they served five hundred million banking customers. To be approached by this company meant that Ohpen was doing the seemingly impossible—and impressively. Everything they were doing was "new"—programming a whole new bank, hosting that bank on the cloud, and offering it as SaaS. It seemed to Chris that this large company wanted to kick the tires and look underneath the hood. "Let's see what these guys want."

At the outset it seemed like a promising deal. Chris, recalling the reasons this company had approached him, says, "They thought that strategically it would be smart to have our knowledge and new software and that we could help them put pieces of their software into the cloud and run it as SaaS." However, it wasn't just a one-sided deal. Chris thought, from a sales perspective, that a takeover would be advantageous to Ohpen. This company already had very good relationships with some thousand banks around the world. "If at least one percent would be interested in our software, we could already have ten new clients," Chris adds. "Besides, I like to make technology. I don't like all these other things. If we could concentrate on making technology and they could sell it because they already have a thousand banks as clients, it would be a good deal."

After everything Chris had been through to secure a client, it's hardly shocking that he would have been willing to partner (at the very least) or sell with or to a competitor. At the same time, however, Chris and Matthijs were also nearly closing another two deals with potential clients in both the Netherlands and the UK. But the UK client wasn't just any client—it was a major investment manager, M&G. It would put Ohpen on the international map. "It was just a very, very hectic time—running this company, getting new clients, building it, hiring people, and serving our new client, Robeco, while selling the company at the same time," Chris recalls.

Seemingly almost overnight, Chris and Matthijs were in the middle of two potentially huge multi-million-dollar deals, and both were being negotiated on the same weekend. They had to be at a meeting in Athens on Sunday to meet the company interested in buying them, and they were supposed to meet M&G in London on the following Monday morning at 9:30 a.m. The only way to pull this off was with the help of their investor, Alex, who lent them his private jet so the two men could be in two places in a short amount of time.

Chris and Matthijs met on the tarmac to board the three-hour flight to Athens early Sunday in July 2013. Chris couldn't get over the contrast. For years he had been living out of airport lobbies, and now he was being whisked away via private jet to close a megadeal. "It was so surreal for both of us. Flying just the two of us with the pilots. And landing with a car waiting for us. It was all so strange," Chris remembers. While on the flight, they worked on the final touches of their presentation. Prior to the meeting, Ohpen and the buyer had already made an informal agreement: Chris was prepared to sell the company for tens of millions of euros. If this deal went through, Chris and the founders would be set for life.

It was a whirlwind day and evening. Chris remembers the car picking them up at the airport and taking them to meet with the whole board of the company to make their presentation. Afterward, the entire board wined and dined them—taking them to dinner in a harbor. After dinner, Chris and Matthijs went back to their hotel and slept for two hours before going back to the airstrip by 3:00 a.m., where the plane was waiting for them. They tried to sleep during the four-hour flight to London to be prepared for their negotiation with M&G, but they were too excited. They also knew they had to be prepared for their next meeting. M&G required their full attention too.

Matthijs and Chris made it to the meeting in time, and within thirty-six hours of their brief adventure, they were back at home and back to business as usual, running Ohpen.

Over the next few months, negotiations with M&G and the potential buyer continued. But something wasn't sitting well with Chris. "There are always these little tricks that companies pull to make changes to the original agreement," Chris adds. Chris had trouble digesting this. "I could never understand this. A deal is a deal. It felt like being stabbed in the heart. I know others say, 'Dude, it's just business.' But it's not for me. At the time, I started thinking that maybe I wasn't made for this." But what one company in particular didn't know was that as a provider to the financial industry, all telephone conversations were recorded. Chris wanted to make sure everyone at the buyer's company was holding up their end of the bargain. This was his company he was negotiating, after all, and he wasn't about to take it lightly. The recording provided the backup he needed, but in reality he didn't need it. Chris has a memory like a steel trap for these kinds of things, and it wasn't hard for him to remember certain deliverables. Chris remembers, "We would sign the agreement on the fifteenth, and we would get the money on the twenty-sixth, two days before my wedding. I would get five million euros cash myself as a first payment. So there was one payment of five million paid on the twenty-sixth for me and the share purchase agreement (SPA) signed on the fifteenth of September, which was my birthday. The most important thing, though, was that I wanted to be able to sell our software to their clients, and I wanted that in writing. I needed the power in the company over the sales force, and if I did not have anything in writing, I was afraid I could not sell our software to their clients." Chris began to worry that if no one was willing to put anything in writing, then perhaps they wanted to buy Ohpen just to kill it.

Then he received calls from the buyer, and the dates, the amounts, and the means of payment changed. Instead of cash, they said they would pay him in stocks, with a payout after a year's time—with the stipulation that he couldn't sell the stocks for three years. Chris called the CFO and said, "Listen, this is not what we agreed on." For Chris, it wasn't about the earnout. He really wanted to be purchased by such a major player, and he wanted the Ohpen software to go worldwide. However, the fact that they'd changed the original agreement made him uneasy. They could have just said that they'd thought about it and wanted to change it. In Chris's mind, they should have been honest about it. But they countered and said they never agreed to anything and that Chris must have heard it wrong. Chris knew this was going to happen, so he refreshed their memories and repeated back the terms of the agreement that he had repeated many times just to confirm when they'd first spoken. After many, many rounds of "we never said that," Chris was stunned by the ease of which the buyers lied to him. He knew he couldn't work with them—no matter how much money was on the line. He could have made a lot of money on a major company's board of directors, but in the end, that wasn't the most important thing. He did tell the CFO that he would send him the tapes so he could hear his own words, and from that moment on, their memories started to come back all of a sudden.

Later that evening, the CEO of the company called Chris on his mobile. Chris wanted him to agree to the original deal—first payment in cash, Chris and team would stay for five years, and they would get an earnout in stocks. That was the deal, take it or leave it. It was a very intense time. Without mincing words, Chris recalls the crucial moment in Ohpen's history: "It was a mind fuck. *Am I going to sell, or am I not going to sell?*" Chris was at a crossroads. He could potentially make a lot of money—just three years into the company—and at the same time have access to unlimited distribution so he and his team could focus on building software, *or* he could back out of the deal and build the company's client base on his own. He knew this would be a difficult path. However, something deep inside was gnawing at him—he couldn't put his finger on it, but something felt very "off," and Chris was a man who followed his instincts. He liked the CEO very much, but the CFO's constant lying was very annoying.

While his intuition was telling him to pull out of the deal, something happened that made him *know* he had to pull out for sure. They called Chris and said that the earnout should only be for Chris and his fellow founders, *not for his investors at Amerborgh.* They asked Chris to help them with this—by offering Chris and the team some extra cash. Chris responded by restating the conundrum they were putting him in, in plain, crude English: "You're asking me to fuck over my investor—the guys who trusted me and believed in me when no one else did?"

Chris was in disbelief. He then proceeded to tell them: "Michel and Alex gave me three million euros, no questions asked. They trusted me—*this person. Me.* I wouldn't steal five cents from him. I would never, ever treat them badly—he and his entire family can always count on me." Chris was incensed. "We've had all these talks, and you have no fucking clue who I am, what I stand for as a person. The thing that you ask, what you propose to me, I could never do. Even if you give me a hundred million. I'm just not that kind of person." Quoting the Mexican revolutionary Emiliano Zapata, Chris added, "I'd rather die on my feet than live on my knees."

The thought of screwing over his investors was too much for Chris and told him everything he needed to know. Chris first had to break the news to his team members, who were counting on the sale—it would have made them wealthy as well. He gathered everyone together, the founders and the investors, and said, "Listen. I don't need my earnout. So if you guys want to sell, that's fine. But *I* won't work for these guys. I'll play the part, and when everything is signed and paid, I'll get into a fight within three months, and they'll fire me. I'm not going to work for these people. And if all you guys want to sell, then we'll do it, but I'll be gone very soon after."

Chris made it clear where he stood. He didn't want to screw his friends out of their potential earnings, but he didn't want to work for these people. In the end, all the founders agreed with Chris and decided not to sell. Not one of them needed convincing, either. All were in complete alignment with Chris.

In the end, the founders put their faith in Chris—in themselves. They didn't need the buyer. They could do this. One of the best parts of the process, Chris points out, was the journey that got everyone to this decision. He and Matthijs were traveling the world, pitching companies, and heard the offer, "We want to buy you." It's not that it legitimized them—they already knew they were the real deal—but it certainly felt like a solid endorsement for their product and services. And Chris adds, "Being at all these meetings with Matthijs and doing what we were doing and selling the company for a lot of money before we even got fucking started was really cool." Chris adds, "The best part was that we went from being the *crazy guys with their stupid idea of going in the cloud* to being *visionaries* in just two years. After this first offer, many others followed from big core banking firms, but we decided not to sell and try to make it on our own."

Besides the journey and feeling like they had arrived, the experience also proved to be valuable in other ways as well. It showed everyone—clients, potential buyers, and even employees—what Ohpen stands for. Ohpen wasn't anyone's fool, and they weren't going to compromise their own principles, values, or integrity just to make a buck. "It's karma. I believe that if you do good, good things will happen," Chris says. Of course, the converse is true as well, so Chris likes to avoid doing things that would in any way compromise his own integrity or the integrity of the company.

> **Ohpen wasn't anyone's fool, and they weren't going to compromise their own principles, values, or integrity just to make a buck.**

And this is why it was all the more disappointing when others didn't extend the same courtesy.

ANOTHER DEAL FALLS APART, AND A VALUABLE LESSON IS LEARNED

While Chris and Matthijs were busy negotiating the potential sale of their company, it was still business as usual in the Ohpen office. They were still going after prospective clients, and they were hoping to extend their reach beyond the Netherlands and had their sights set on the UK. However, after months of negotiating a deal with their UK client, M&G, they were close to signing. All the client needed, they said, was their board's approval. It was as good as done. Chris, Matthijs, and the team had no reason to doubt them. They emphatically declared several times, "We have the contract, and we agree with the contract; we just need our board meeting to happen, and then we'll sign it. Can we already start?"

So, on good faith, Chris and Matthijs gave the team the go-ahead—they could start working on onboarding the new UK client, M&G. For the next two months, the Ohpen team operated on good faith, pouring money, time, and manpower into the project. Erik had already pulled his children out of their school and began preparing for a move to London. Lydia was already searching for an apartment in London as well. Then they heard the news: the board was going for another option. They weren't going to be able to sign the contract after all.

"This was the first time I saw Matthijs pissed off in the fifteen years I'd known him," Chris remembers. Chris thought, reasonably, that the M&G guys would at least compensate Ohpen for the work already completed. But the firm balked and said, "That was all presales work; you're not getting five pounds from us."

Matthijs had worked tirelessly on this negotiation for months. Matthijs, who was normally calm and unflappable, expressed his incredulity: "We spent over two thousand hours of work, spending two hundred and fifty thousand euros out of our pocket, and you're not going to sign, when I've met the full board and obtained their verbal approval?" M&G responded that it had all been clearly stated as "subject to board approval."

Chris says, "If they thought the risk was too high, I could understand. If they actually did not get approval from someone higher up the ranks, I could understand that as well. I rationally can understand why those big companies could not sign with us. It's all good. These things happen in business. But having us start the project and work for months without pay and not even having the decency to pay our hours or even a portion of them? I just can't understand how anyone can do this. Maybe I am not made for this, but it is mind blowing how those board members of M&G just sat there and did not want to pay us even one pound."

Matthijs wasn't the only one pissed off. Chris could hardly contain his own ire and recalls, "I came very close to practicing some martial arts on that dude sitting across from us delivering the news with that irksome smile and not giving a fuck about us."

As mind blowing as it was, it was yet another valuable lesson learned. While Ohpen had a deep sense of what was right and wrong and standing by their word, they couldn't expect everyone to share the same values. They also chalked the experience up to a lesson in cultural differences. The Dutch are known for their direct and straightforward approach—telling others exactly where they stand and expecting the same in return. However, the Brits, while overly polite (in other words, less direct), were more difficult to get a solid read on. They could be saying everything is "splendid" but thinking something altogether different. No matter—the lesson was learned. Henceforward they would get the contract signed before a dime was spent. Chris reflects about the difference between the Brits versus the

Dutch: "The Dutch are too honest to be decent, and the Brits are too decent to be honest … so we should have known."

DOING THEIR HOMEWORK

In addition to being value driven and focusing on integrity when dealing with both clients and potential buyers and partners, the team at Ohpen relied heavily on their committed values—to give it all, do more with less, and exceed expectations, even when it came to other sharks in the water. And one of the biggest sharks sinking its teeth into the entire banking world was the US's Foreign Account Tax Compliance Act (FATCA). This was a 2010 US federal law requiring all non-US foreign financial institutions (FFIs) (or foreign banks) to search their records for customers who are US citizens. In other words, the US had instituted a law that reached beyond its own borders and required *all foreign banks* to comply. This meant the software that banks were using needed to be able to search individual records that stated US place of birth, prior residency, or similar, and then had to report the assets and identities of such persons to the US Department of the Treasury. (And as a result, many banks decided not to accept US citizens as clients for this reason.) FATCA also required US citizens or Green Card holders living in other countries to self-report their non-US financial assets annually to the IRS. This was an enormously costly endeavor. Jane Gravelle, a specialist in economic policy at the Congressional Research Service, in her 2015 report *Tax Havens: International Tax Avoidance and Evasion*, discovered that to enact such an initiative costs banks *$40 billion*; however, the actual additional tax income for the US is far less each year—around $250 million.[1] Despite its pointlessness, the law was the law, and European banks took it seriously, because in order to enforce the compliance of banks, the US threatened punitive withholding levies. Needless to say, banks were taking it very seriously, and so Ohpen needed to as well in order to assure their clients that they would be able to comply with the complicated tax code.

In order to implement these changes, Matthijs asked Kees Postma (who Matthijs dubbed "the youngest and brightest guy on the team") to create a functional design that would be compliant with the tax code and would be ratified by Erik and Matthijs. Kees had been an intern at Ohpen (twice) during his university studies and was present for the migration of Robeco. A bit of an outsider, Kees was unlike the guys who dropped out of college. Rather, he was a scholar through and through. But he didn't let his cerebral/scholarly side show too much when he first started. "I caught on quickly and saw what they *really* valued," Kees says, remembering his early months at Ohpen. "I came from a very privileged background, so it was very easy for people to think, *Oh, this guy has had everything handed to him on a golden platter.* I kind of felt like I had to go the extra mile to prove like, *No, no. I'm not one of those people. I'm actually here to do the work and do not just expect everything to be handed to me.*" Kees wanted to show everyone he, too, was willing to roll his sleeves up and do the work. He worked from eight in the morning until midnight (and often longer) to prove he was just as dedicated to the Robeco project as everyone else. And when it came to FATCA, he knew what was on the line for Robeco. The "or else" ultimatum the US was threatening was a 25 percent tax on all proceeds out of the US toward the bank's investment fund.

When Matthijs handed him the project, Kees ran with it. "I just printed out four hundred pages of the law and just started reading it, understanding it, and I just kind of …" he says modestly, "internalized the whole thing." In fact, he had basically memorized the entire technical design requirements. When Robeco came with their

FATCA specialists to talk about the impact the code would have and how it should be implemented in the software, Kees says, "I just knew it inside out. I could point to the exact page and name the exact code of whatever they were referring to." Kees was by all accounts the youngest person in the room by ten years that day. When Robeco threw up an objection to a design, Matthijs remembers seeing young Kees clear his throat and reply, "Excuse me, but according to section X, sub Y, on page Z, it clearly states that …" It was a moment of particular pride for Matthijs to see Kees, ever the scholar, also in that moment become the consummate Ohpen employee—direct, smart, determined, willing to give it all. Most importantly, Kees had rolled up his sleeves and had done his homework.

These experiences—dealing with a potential buyer, being burned by a potential client, and going up against a behemoth tax code—though challenging, showed people within and outside of Ohpen what Ohpen's values were. They stood by their word, they were loyal, they were transparent, they always gave it their all—and they always did their homework. They were nothing if not thorough and competent, and they took their clients' compliance seriously. In fact, they took their clients, compliance, safety, and security more seriously than anything.

> They stood by their word, they were loyal, they were transparent, they always gave it their all— and they always did their homework.

CHAPTER 10

ENSURING SECURITY

DESPITE THE DISAPPOINTMENT with M&G, Chris and Matthijs continued to meet executive board members and leaders of banks to show them the benefits of the Ohpen platform. From the beginning, it wasn't easy. Despite all evidence to the contrary, many people still had a hard time being convinced that the cloud was a perfectly safe place to store data. Chris recalls, early on, one COO of a very large insurance company turning to him and saying, "The cloud is not safe." This wasn't the first time Chris had heard such an objection, and he also knew that most of these objections come from people who don't even understand what the cloud is, so Chris responded by asking just that: "Well, what is the cloud? Can you explain it to me?" The COO shook his head and just repeated to Chris the often-heard reply, "It's just not safe." Chris tried to be patient and responded, "When something is new, it's normal that people don't understand what it is. That is human nature. I'd be more than happy to explain what it is; it is really something amazing. Let me explain it, please. You'll see," and he proceeded to do so. Chris heard such responses over and over, and it bothered him that people making decisions for banks did so not based on fact but on a fear of the unknown. To Bas it felt almost as if they were cowboys. "We might as well have pulled up with horses outside," he reflected. Chris felt that selling SaaS on the cloud was almost like inventing a car and selling it to people who just wanted a faster horse. They couldn't possibly conceive of hundreds of horses or massive horsepower for a price cheaper than the one horse they owned. "It almost drove me crazy that people did not understand," Chris remembers.

While what Ohpen was offering was new, it wasn't entirely unheard of at the time. In fact, around the same time that Bas, Ilco, Erik, and Chris were building the platform and planning to put it on the cloud, the Dutch Central Bank sent out a circular stating that cloud computing would be treated exactly the same as outsourcing. Outsourcing was already addressed in the Dutch laws. So Chris's ideas wouldn't just have to prove it was safe but that it was compliant as well.

RISK MANAGEMENT AND THE SAFETY OF THE CLOUD

According to the law, banks can't outsource without doing a proper risk assessment. This, Matthijs adds, is completely understandable. "When making important decisions in general, one wants to understand what one is signing up for—not only looking at the opportunities but also the risks involved. The PRU calls this kind of decision-making 'integral company operations and controlled decision-making.'" In other words, banks need to be able to demonstrate that they can make decisions the right way, and the PRU needs to be informed before a decision is executed. The law demands that banks create a risk assessment before making a decision and sharing that document with the regulator. In order for banks to know what is expected of them and to help them move in the right direction, the Dutch Central Bank proposes that banks use a template that in turn is based on the European Union Agency for Network and Information Security (ENISA) framework. ENISA, which comprises subject matter experts from various technologies industries (among them Google, Microsoft, IBM, and many more), academia, and governmental organizations, created a risk assessment for an on-cloud computing business model and technologies and provided practical recommendations for use. All banks are incentivized to use this framework, and they outline what ENISA defines as the top security risks. According to the ENISA white paper *Cloud Computing: Benefits, Risks, and Recommendations*, these are the major threats: loss of governance, lock-in, isolation failure, compliance risks, management interface compromises, data protection, insecure or incomplete data deletion, and malicious insiders.[2]

ENISA distinguishes about forty questions (with some subquestions) that are further broken down by organizational risks, technical risks, compliance risks, and non-cloud-specific risks. Ohpen executed this very assessment before entering into a relationship with AWS, before taking their platform on the cloud (and they still do it for any major outsourcing agreement). Matthijs adds, "We also assist our clients in executing them."

Over the years, Matthijs has become Ohpen's de facto expert on risk management, and he understands that one can never *exclude* risks completely. "One would never achieve anything, let alone create value for the company or its clients, if no risk is ever taken," Matthijs says. Therefore, he and the Ohpen team identify the relevant risks, find the best mitigating measures to prevent the risk from materializing, and determine what they will do when the *net risk* actually happens. "It is all about the *likelihood* of materialization of the risk and the *impact* it has when it does," Matthijs says. "A good risk assessment tries to cover the entire field of possibilities but focuses on what really matters. Multiple disciplines must be involved, because it is in the interaction between different specialisms that certain risks are identified—and that mitigating measures are thought of and are implemented."

ORGANIZATIONAL RISKS

The first part of the assessment is related to organizational challenges. For example, *how does one make contractual arrangements with the cloud vendor?* Matthijs likens the contract to a marriage. He asks, "How do you make sure you know each other well enough before you engage in marriage? And the less easy topic: How do you make sure you can part ways if you feel it is better for your company?" In other words, how does a company make sure they are not *locked in*— which is one of the major threats that ENISA warns against. "An outsourcing relationship shouldn't be Hotel California," Matthijs jokes. That is why a bank must create an exit plan even before entering into the relationship. "Other things to consider," Matthijs states, "are what happens if the company you outsource to is taken over? And how do you make sure that all throughout the value chain, rules are respected and that your provider can count on its subcontractors to deliver?" Another risk that is identified by the regulator that Matthijs and Ohpen take seriously is that when they outsource—whether to the cloud or not is irrelevant—they risk losing knowledge of mission-critical processes and they won't be able to challenge the provider. For example, if you outsource reconciliation of cash transactions and stock market trading but you have lost knowledge of how these processes should work, you can never properly assess whether your outsourcing partner does it correctly.

TECHNICAL RISKS

In addition to organizational risks, there are inherent technical risks when subcontracting. For example, *can a bank ensure that the provider can handle its volumes? Can it ensure that it doesn't mix a bank's data with data of other clients or lose control of the integrity of the bank's customers' data?* Another question to ask is this: *How is access allowed only to those who should have access, and how are identities verified?* Data both at storage locations or in transit over the internet or VPNs is at risk of being intercepted. So then, how are systems *hardened* to limit the possibility of being tampered with? *How can one make sure that the encryption keys aren't lost?* Is there a high risk of any distributed denial of service?

COMPLIANCE RISKS

Matthijs notes, "Being compliant when you do everything yourself is already quite the challenge." In addition to maintaining one's own compliance, banks have to make sure their compliance requirements are met when they outsource as well. Matthijs adds that there are several factors to consider: *Is there a right to audit? Where is the data stored? What happens when one deals with different jurisdictions that have conflicting regulations?*

Additionally, apart from IT risks that are specific for the cloud, numerous risks exist that are not cloud specific. There is everything from theft, natural disasters, conflicts of interest, unauthorized access to premises, backups that are lost or even stolen, and everything in between that must be considered, no matter how unlikely that it would happen.

The final note in the ENISA white paper on potential risks states that a bank can "outsource responsibility, but it can't outsource accountability."[3] In other words, the burden of risk falls on the *banks*. This means that the banks must verify whether or not the outsourced company meets all of its accreditation guidelines, such as ISO, SOC1/ ISAE 3402, and SOC2, as well as assess their level of potential risk. Ultimately banks assess the outsourced companies (also according to ENISA's guidelines and recommendations) to make sure they have fully mitigated the potential threats to security. Because at the end of the day, it's the banks who hired the outsourced companies that have to answer to the DNB. And the DNB's ultimate job is to make sure that banks are regulated and follow the law in order to protect consumers. The right to audit is the strongest tool in their toolbox to do this. The DNB, however, is not the only one that has the right to audit. As mentioned earlier, banks also have the right to audit. In fact, it is in their best interest to do so. For it is the hope that auditing, when done properly and consistently, prevents the DNB from ever needing to exercise their right to audit. (The DNB is usually called in when there is a serious problem, and most banks want to avoid this at all costs.) In order to prevent such serious problems, banks will do a regular audit of an outsourced company—to assess both that accreditation rules are followed and whether everyone in the company follows the procedures written in the company's own security and safety policies. An extreme example Chris likes to give is that if a company wrote down in its procedures that everyone has to open the door with his or her left hand, then an audit would be done to check that everyone in the company complied with that rule. "The rules don't have to make sense, but they do have to be followed once they're written down," Chris explained. However, this is not the only type of audit being conducted. In essence, there are two types of audits—one is conducted to make sure that an outsourced company does what they say they're going to do (i.e., opening the door with the left hand), and the second type of audit is when the bank has an opinion *on how* something is being done. In other words, banks have a right to comment on what procedures should be followed as well as how they should be followed.

COMPLIANCE WITH INTERNATIONAL FRAMEWORKS

In addition to a company's policies, procedures, and controls, outsourcing companies need to comply with international frameworks. Matching and setting internal safety and risk protocols to the international framework was one thing, but Chris, Matthijs, and the team also had to take it a step further and started to assess their clients' (the banks') policies as well. Every bank has its own security, risk, and data protection policies, which Ohpen would have to comply with in order to do business with them. A bank can say, "Show me your encryption policy," or "Show me your identity and access management policy," and Ohpen must comply. They can also say, and they do, that Ohpen has to comply with the bank's encryption policy, even if it is different from the one Ohpen uses. Chris never saw this as a bad thing. Always looking further down the road, he thought it would be a good idea to comply with all the policies of his existing clients. In his mind, it would be a massive selling point to Tier 1 banks to say they are compliant with all the procedures of all clients.

It all starts with the law. The DNB will interpret the law as agreed by parliament and sanctioned by the senate, make guidelines, and judge the risk assessments of the bank, but the three are actually governed by one umbrella framework. In addition to the policies, international guidelines, and accreditations, the DNB has the right to audit companies that banks outsource to. So that means that if the DNB suspects fraud or some violation in a bank, they have the right to audit not just the bank but all the outsourced services to vendors, including the cloud host—yes, even AWS. Anything that is outsourced is still a bank's legal responsibility. So, of course, banks take the security and safety procedures of any outsourced company very seriously. "It starts at the law," Chris says. "First, if banks outsource, they are responsible as if they were executing the tasks themselves. They can never outsource compliance responsibilities. Second, if they outsource, they need to do a risk assessment and inform the prudential regulator of the outcome. Third, if they outsource, they need to have an audit right and make sure the regulator has no barriers to overcome when executing their duties. Those are the three basic rules from the law."

This is all to say that when Chris started Ohpen, he was well aware of the inherent risks of banking and of ensuring the safety, security, and privacy of their clients as well as the compliance proce-dures and risk assessments. Knowing this from the inside out, Ohpen built a system with all of this in mind from the ground up.

WHERE THE DATA IS STORED

Bas knew from the beginning that the first thing he and Chris would have to convince every bank that wanted to do business with them was that their software was safe and, moreover, that where it sits—on the cloud—is safe. "The first question we always get when I'm sitting with a new potential client is, 'Will my data reside in the EU?' Because as soon as it's stored in the US, they worry that the US government will have access to the data through the widely inter-preted Patriot Act and the USA Freedom Act. This is not allowed by European data protection regulations," Bas says. "But the reason we love AWS is they have divided the world into regions, and the data will never leave the specified region unless a bank requests it them-selves." In fact, according to the AWS, they have servers located in "sixty-four different availability zones" within twenty-one geographic regions around the world.[4]

What this means for banks is that they retain *complete control over their data* in the region their data physically resides. In other words, just because AWS is an American company, it doesn't mean the servers are located in the US. AWS even created a European sub-sidiary in Luxemburg to make sure European clients can comply with local regulations and that this AWS subsidiary itself is governed by EU regulations, not US ones. In addition, within regional availability zones, banks can choose to increase redundancy and fault tolerance by replicating data in various regions—something legacy systems can't do. AWS also provides more security and privacy controls than most banks can themselves. It has network firewalls and web application firewall capabilities that allow core banking engines like Ohpen to create private networks and control settings. It also includes state-of-the-art encryption. And unlike legacy systems that rely on a patchwork of upgrades, AWS was built from the ground up to provide what AWS calls "resilience in the face of Distributed Denial-of-Service (DDoS) attacks." Current legacy systems weren't built with the internet in mind and could not conceive of the sophisticated DDoS attacks that are commonplace on the web today. Cybercriminals can now send malicious bots that disrupt normal traffic on a targeted server, overwhelm the target and its surrounding infrastructure, and flood it with internet traffic and faulty log-on attempts, ultimately blocking users from using the servers. Bad news for banks with legacy servers. In addition to their ability to avoid and detect attacks, AWS's auto-scaling features make it nearly impossible to disrupt traffic increases. In addition, AWS offers layers upon layers of security—including the ability for key management—something Ohpen values. Key management allows for Ohpen to control the encryption keys. So AWS servers may have encrypted data, but Ohpen holds the encrypted keys to that data. In other words, the two pieces of the same puzzle are not stored in the same puzzle box. In one "box," a disk or data in a relational database is encrypted and stored, and the key to decrypt the data/disc is in a completely different "box." The other half of the answer to the question lies elsewhere, making sure that only the lawful recipient of the decrypted data knows where the "key" is and therefore can read the outcome. Unlike most legacy systems, AWS is constantly monitoring and looking for threats. It automatically assesses systems for what they classify as "vulnerabilities and deviations from best practices" and automates alerts of impacted networks. There are also tools that help IT teams spot issues before they impact the business and help reduce risk and increase the security of the environment. No current legacy system can do *all of this*. Needless to say, when a CEO or COO of a bank now says to Chris or Bas, "The cloud's not safe," they may want to take a seat, as Bas says, "This could take a while to explain."

ENCRYPTION

However, Bas argues, *where* the data sits should be the least of a bank's worries. The most important security measure any bank can take is encryption. "If you are able to encrypt the shit out of everything and you are the one who controls the encryption keys, you can answer almost any question of safety. And that is the truth," Bas argues. "So a customer of a bank logs in, and it goes through at least three or four levels of encryption. The connection is encrypted, the database is encrypted, the disc is encrypted, and the passwords are encrypted—all before somebody's logged in. *And that's every time.* And the reason behind it—I always say that three is the magic number—is if one protection mechanism fails, like security of the disc, we still have two or three left." With so much encrypted data, there is a point at which the data needs to be accessed. There is data at rest (which is stored) and data in transit. In both cases, the data needs to be encrypted, and in both cases, Ohpen needs to make sure the data is completely safe from hacking. "We didn't want to use some commercial-grade thing made by some Chinese company with some backdoor stuff we couldn't control," Bas adds. "We have to be able to prove that what we use for encryption is solid and proven to have no back doors."

Raymond Morsman, the senior security officer for Ohpen, who was also a former BinckBank employee, spends his days maintaining Ohpen's security. In his own words, he says, "My daily job basically consists of making sure that all the controls we've put in place to get our SOC2 and ISO are met and that everybody is doing what they're supposed to do with those controls." Though his job to ensure security is consistent, his day-to-day job varies depending on what needs to be attended to. One of the things he does is run scripts that connect to a built-in vulnerability scanner. The scanner automatically shows him any potentially vulnerable areas that arise from seemingly innocuous software updates. "Vulnerabilities exist everywhere," Raymond asserts. "They can be in Windows updates that need to be run. If you have a tool installed like WinZip or Word, all those software packages get updated by their suppliers."

Raymond is passionate about security and sets aside time each day to read upward of five or six articles related to the latest updates to security as well as potential hacks and threats. One area of particular interest to him is encryption. "I am always looking for the latest news, upgrades, and best practices," Raymond says. "Because encryption is the most important part. Your keys and your data are important. The whole reason we started using encryption, even before we had a client, was because we didn't want anyone ever to get their hands on our data. Imagine having a machine and completely encasing it and protecting it with firewalls and then putting your data on there without encryption. If you don't encrypt it, that data can suddenly pop up somewhere else, and that's a huge problem. It's impossible to recover encrypted data as long as you use the proper level of encryption. When we terminate machines, we want to make sure that those discs that contain data are not readable anyway."

It's not just data at rest that is encrypted, Raymond adds. All the data traffic between machines is encrypted as well in order to protect from malicious insiders. "We want to make sure that data we send off the line is just as secure as the data on the disc. So all traffic that we send through AWS is always encrypted. It's quite easy, actually, because most programs nowadays support DLS (Distributed Link

Service). And that's what we're using for all the communication between the machines."

In addition to all encrypted data at rest and in transit, Ohpen built its own software to keep the keys to that encryption. However, they use Microsoft's BitLocker as their encryption software because, Raymond says, "It ticked all the boxes that we wanted." Another benefit of using BitLocker, as Raymond sees it, is that "it is basically guaranteed to keep working, because Microsoft will never just abandon their encryption protocols. They might update them and make them better, but they will never abandon them." A further advantage to using BitLocker to encrypt data is that it is widely considered "unbreakable." Former encryption programs such as DES, which was considered "secure" up until the late nineties, most hackers can break into under twenty minutes. BitLocker at this time is virtually hack-proof—at least from an encryption standpoint.

In addition to encrypting all the data with state-of-the-art encryption software, Ohpen also monitors all the data—at rest and in transit—on a continuous basis. "We are always monitoring all traffic to and from our environment at AWS. So it's not like, 'Here, I got some encryption, and I don't have any responsibility'—we have to maintain it continuously. *And* we have to keep up with the latest encryption standards—what is supported, what is not supported. Many times, we knew something was wrong with our clients before they knew it themselves," Chris explains.

According to Bas, "It was way easier to implement encryption in a solid way than to find all kinds of ways to show the Dutch Central Bank that we're in control of the data. Because the starting point is, we are a data *processor*. We don't *own* the data of our clients—the banks do. If you look at privacy laws, there's a clear distinction between the data processor and a data controller." By definition, the data controller is the person (or business) who determines the purposes for which, and the way in which, personal data is processed. By contrast, a data processor is anyone who processes personal data on behalf of the data controller (excluding the data controller's own employees).[5] "So that means that we're bound to certain rules. And that's good," Bas adds. Bas and the team believe these rigorous rules and standards benefit not only themselves but the entire industry. "You have to imagine it's like open-heart surgery for these banks to give us their personal data, all their customer data, and say, 'Here you go; we trust you—don't misuse it. We trust that you have qualified personnel and that you can do whatever is necessary to process the data.'"

Knowing the seriousness of data protection, the Ohpen team built a fully integrated safety, security, and privacy policy to ensure that all customer data is protected all the time—not just from DDoS attacks but from hackers with other malicious intent as well. While Bas and the team are confident in their security protocols and encryption methods, they are fully aware of the potential threats outside their control. With hundreds of billions of internet-capable devices in use today—computers, printers, laptops, phones, even refrigerators and monitoring devices—each device out there is potentially hackable. (And most clients don't know that printers are the biggest weakness in a company's security.)

SHARED RESPONSIBILITY MODEL

That is why the shared responsibility model is so important to the work Ohpen does. AWS is responsible for everything that is on the cloud, or what they define as "hardware, software, networking, and facilities that run AWS cloud services."[6] They are responsible for monitoring potential hackers and activity on the cloud constantly. However, in the shared responsibility model, the end user is also responsible for compliance and security. So it's on Ohpen to encrypt and monitor—and manage data. In other words, people are a huge part of the shared responsibility model. Chris, Bas, and the entire team take this extremely seriously. "The shared responsibility model has been, from day one, something that we just drilled our employees to get knowledge about." Basically, the shared responsibility model assumes the responsibility and management of the guest operating system (including updates and security patches) and other associated application software as well as the configuration of the AWS-provided security group firewall.

ENSURING SECURITY INSIDE OHPEN

Educating the employees and maintaining safety as top of mind is a foundational procedure at Ohpen. "It's quite simple. Our job on a daily basis is not only to assure clients that their data is safe but to make sure that every employee here knows how to keep that data safe," Bas says.

All the safety procedures within Ohpen, from entering the building through a circle lock and being scanned and weighed upon arrival and exit to fingerprint access to different areas within the office and video surveillance throughout the building, are the bare minimum. Monitoring of *all* the devices to ensure the protection of data and limit the access to servers is even more important. The authorization matrix that is applied to the personnel allows Ohpen to demonstrate to their clients that only the correct people have access to data or servers.

Security begins even before an employee is hired. "Everybody who starts working here is carefully screened. In addition, all employees who have access to production data go through rigorous background checks, including criminal and financial records, and are continually screened multiple times a year. This includes credit checks to make sure no one is bribable. There are also periodic drug and alcohol screenings to make sure everyone at work is working at optimal capacity. It's that serious," Bas says.

In addition to making sure the data is encrypted, the people who work at Ohpen, especially those operating the systems, are subject to several security measures. Ohpen leaves nothing to chance. "We don't trust any network other than our internal network," Bas says. "This can be done because of the cloud. We call it the *cattle versus pet* method. Most companies treat their applications and systems like pets. They coddle them, and if they're sick, they try to fix them or patch them and try to make them work. We treat our systems like cattle—if they are causing trouble or they can't be fixed, we just shoot them. If we have ten machines running fine and one is not working fine, we just kill the machine and launch a new one. And that's a totally different approach—managing an infrastructure without trying to fix stuff. If stuff doesn't work, we just kill it, because we have other machines to use where things are working fine." In addition to isolating or killing broken systems, Ohpen does everything possible to isolate traffic that comes in and out of Ohpen.

All the engineers who work in system operations, when they are on their computer screens working, have no access to the internet—that means no Google, no Facebook, and no email—so zero threat that an email with a faulty link can be sent to them. And they all have separate accounts, so they can be individually monitored—when and where they sign in from. And systems are automated and can alert managers when unauthorized log-ins are being made. For those employees with laptops or company phones, all data can be swiped and removed from anywhere in the world at any time in case an employee has lost their device or is let go in an unforeseen manner. Nothing is ever left to chance.

However, cases that require these protocols are few and far between, in large part due to the extensive and thorough onboarding process. Every employee—no matter what level or department—goes through intense safety and security training. In addition to an online security awareness educational program, employees are expected to abide by the Ohpen code of conduct as well as the rules and regulations as laid out in the personnel handbook. In addition to training, Bas makes sure that all of Ohpen's employees are compliant with the Service Organization Control (SOC) Reports (1 and 2). The SOC1, or ISAE 3402, manages all of the core processes in the application that have or could have impact on the client's financial records, and it makes sure that all the processes that Ohpen has outlined and written down in the policies and procedures manual are actually followed to the letter. "What we try to do with being certified is have those accreditations to show that we're in control of what we do and, in the end, of the data of our customers. We have over four hundred items for which we can prove that we've written down a policy and a procedure, and we can prove that we really work this way." After ten years of building out security and compliance measures, the Ohpen team sometimes finds things on their clients' networks before they do. "It is funny sometimes when we call our clients and ask them if they could check something on their own network. On the other hand, it says a lot about the maturity of our organization and of our processes," Bas adds.

ENSURING SECURITY OF OHPEN'S VENDORS AND BUSINESS PARTNERS

It's not just their own internal processes and employees that undergo intense screening and monitoring. Ohpen does the same to their own vendors and business partners. "We have controls on how we manage our external providers, like AWS, but also our internet providers," Bas explains. Every single vendor is screened and then managed in order to maintain the company's security. Ohpen's procurement policy is rigorous to make sure that only vendors that comply with all the functional and nonfunctional requirements are selected to engage with Ohpen. And Chris adds, "The cool thing is that we get to work with these large financial institutions that have a lot of knowledge about security and risk as well. We learn from them. I would say we learn from each other. We are able to tap into an enormous amount of knowledge from our customers and make our own processes better and better."

All of these protocols are also managed internally in three ways. "Our first line of defense is the people working in the operations," Bas explains. "Then the second line is our risk department, the compliance department, and my team. We are constantly checking whether the processes are executed properly." And the third line of defense is an internal audit. "The entire purpose of an internal audit is to check to see if everyone is executing all the controls properly," Bas adds.

Chris agrees and decided that the next goal was to automate all internal audits. "Because," he states, "if we've automated all of the controls, we don't need an internal auditor to check on us, because everything is automated and the evidence can't be tampered with."

It's an ambitious plan, Bas concedes, but he and his team are already in the process of completing it. His goal is to ultimately eliminate the need to manually audit. The purpose of doing so, he believes, is to "massively reduce the burden of constantly proving that we're in control." Chris says, "I want to automate the entire second and third line of defense, and we'll get it done soon. Bas and his team will get it done. I think we'll be the first to have done this. Most people find these back-back-back end processes boring, but we find it amazingly cool that we'll have a totally automated robot that does all the checks, balances, and controls for our entire company. We'll just have to watch a screen. I have said that I want the chairs of Joey and Chandler featured in the sitcom *Friends* so our team can just sit in them and watch the screens. When we can do that, then we will have done a good job—when the team does nothing!"

Bas and his team have set up an internal automated dashboard that is essentially a checklist of all the procedures and regulations that are constantly being monitored in real time. Bas and his team can log in and see exactly where issues (nonconformities) are (they appear yellow or red) and where there are none (green). And when a procedure or regulation is considered a nonconformity or followed up on, he and the team are alerted immediately in real time. For example, if the cleaning team is supposed to have their annual background check conducted, the date it is due is flagged. If the procedure is missed, Bas and his team will know about it. If there is a password error—or too many people are using the same key pass—Bas and his team will be notified in real time. If there is ever an issue with the technology, again he and the team will be notified. He is quick to add, "The technology is airtight." Within the automated module that Bas and his team have built to constantly run internal audits are dashboards that monitor their SOC1, SOC2, and ISO compliance as well. Ohpen is constantly being checked for ISO compliance in areas of competence management, skills retraining, awareness, communication, operational planning control, security risk management, risk treatment, and cryptography. In addition, they are checked for security, availability, processing, confidentiality, and privacy by SOC2.

The way the dashboard is set up now, Ohpen is able to monitor, and thereby control, all of their procedures to make sure they are always in compliance. "We're always ready. Even before SOC2 or a client makes a request." Bas adds, "We didn't want to do things double, so we created our own control framework. We just map those controls between those accreditations. And whenever we get a new request for information from a potential client, I can look here on our dashboard and deliver all the information just like that." Chris adds, "It took us ten years, but now we have created something really special. As IT nerds, we sometimes forget to show our clients how awesome the technology we have is. We love building awesome tech like this, but we need to actually show the world that we have it. When our clients finally do see what we have, they always ask us why we haven't shown them a demo of it before. It always gives them a lot of comfort to see it in action."

Ensuring safety and security is perpetually ongoing. It is a continuous improvement process. Every quarter, the Ohpen legal, security, risk, and compliance teams meet and discuss the processes and rate them from green (compliant) to red (not compliant) from change management, release and incident management, and everything in between. "If they're green and everything's under control, we just raise the bar and set them to orange again," Bas adds. "We're constantly looking at ways to make things better. That really means that we're always taking another fresh look at what we're doing and why we're doing it. We have to be in control. It's a never-ending story, but I think the way we're doing it right now is good. I'm not saying it's the only way, but with the number of people we have, the knowledge, and what we try to protect, I think we're doing really well." In addition to the internal Ohpen teams meeting every quarter, the security teams also meet with the clients, and everyone brings something to the table. Chris adds, "Our clients actually make us better and better as well."

The entire team is confident in the security and compliance of their company. After all, they have had ten years of running a bank themselves and ten years of sitting on the other side of the table to perfect their processes. However, it was difficult to convince banks to come around and see that the Ohpen way was a good way that could lead to Rome. As Chris says, "We had been doing that from before the time of the internet and then when it started, matured, and moved into mobile. It's all we have done our entire career. We must know something by now."

Ohpen's knowledge about the cloud and security and their proven success with Robeco as well as their solid reputation was beginning to get noticed by many more banks. Ohpen was ready to scale.

PART FIVE

SCALING
(2014–2019)

CHAPTER 11

CASTING THE NET

ON THE HEELS of successfully migrating Robeco, Chris, Matthijs, and the team were able to enjoy another success. In May of 2013 The Nationale-Nederlanden Bank (NN) began talks with Ohpen about becoming their SaaS provider for all its savings and investment products. From May 2013 all the way through the following spring, NN did extensive due diligence before officially signing the contract in March 2014. One year after Robeco went live, NN and Ohpen signed a second long-term contract. Ohpen now had two clients. They were not a fluke or one-trick pony.

NN MIGRATES TO THE OHPEN PLATFORM

While NN was in the final process of implementing their internet savings accounts onto the Ohpen platform in 2014, Chris thought it was time to pass the baton and hand off leadership of a project to one of his protégés—Jan-Willem. Again, Chris knew if he wanted the company to scale, the leaders and cofounders couldn't be running every project. Even though Jan-Willem was still fairly young and it would be his first implementation, Chris was confident Jan-Willem was up to the task. Chris was there to advise him all along the way. "One of the most important things during the implementation is giving our clients the feeling of comfort that we are in control," Chris says. At the time, he recommended that Jan-Willem visit the NN office three or four times a week to answer their question, show them demos, and help them configure the platform. "You'll be there, and we'll build it here," Chris told Jan-Willem. "Just be visible. Have all the hundreds

> "One of the most important things during the implementation is giving our clients the feeling of comfort that we are in control," Chris says.

of people working there know that *you're* the go-to guy when this big project opens. And he did that."

For two or more days each week, Jan-Willem did just what Chris asked. "I made sure I was at the NN office in Rotterdam, whether alone working there or being out on the floor, giving them the opportunity to talk directly to me. I was the connector, if you will, between Ohpen and NN." As part of the development and implementation plan (DIP), Jan-Willem was in charge of making sure every part of that plan was executed and followed up on. "The first thing I did as a project manager responsible for implementations was to make a timeline of these big milestones and then break things down in further detail." This was all new to Jan-Willem. "Until that moment, I had only been a part of the engineering department teams, and now it was a lot bigger. There's a lot more going on in an implementation than just development." Jan-Willem had to adjust his proverbial lens and pan out and see the big picture for the first time and see how all the teams operate and depend on each other. "The first milestone was to cover all of these things and then detail them. And then I had to plan all these workshops and make sure the teams worked together to deliver all these features—and I had to make sure all of the IT department delivered as well. And then I needed to make sure NN was comfortable and test it." As the single point of contact, Jan-Willem made sure he was constantly in communication with everyone in order to keep the project moving forward. His ultimate goal was to go live at the end of the year and meet their migration goals.

Like Matthijs, Jan-Willem didn't underestimate the importance of face time and building relationships with his clients. He made sure he went to lunch with the NN team or went to their office bar with them. Outside office hours, Jan-Willem made himself available too. "The NN project manager and I went to a bar sometimes to have a beer and talk unofficially about the project. It was all relationship management. I liked that. One of the guys responsible there was a bit older than I was but I think more or less had the same character and interest, so we clicked. That was very important." Jan-Willem credits their good relationship as the reason things got done so efficiently. "He made sure that things I wanted got done internally there, and I did the same, so it was good." At not even thirty years old, Jan-Willem was managing some thirty people and an entire project. Managing people seemed to come easily to Jan-Willem, who has an intuitive and natural leadership style. Charismatic, patient, a good listener, and easygoing, Jan-Willem had no trouble getting the teams aligned. The most important thing in his mind was keeping the project on task. And unlike the Robeco migration, which was carried out over one weekend, the NN project followed a slightly different rollout— they requested a staged migration. First, they rolled out the platform to friends and family, and then there was a soft launch, followed by a hard launch.

After a year leading the implementation team, Jan-Willem was ready to hand the project over to the migration and client management teams. Chris was pleased and says, "Jan-Willem did wonderfully, especially because it was his first time." Chris knew he could trust Jan-Willem to lead going forward.

They migrated NN in April and May of 2016. And while the Ohpen team worked on implementing the Ohpen platform for NN and Chris, Matthijs and Angelique continued to seek out new clients—casting their net wide.

Rfps: WIN SOME; LOSE SOME—
THE ONE THAT GOT AWAY

In October 2014, Ohpen participated in yet another RFP process, this one held by a Dutch bank that was part of an insurance company. This was Angelique's first experience during a formal RFP with Ohpen, and she recalls, "We didn't know anyone at board level at the time." This meant they were at somewhat of a disadvantage going in, because the other competitors did have those relationships. It also meant that their RFP response had to stand out. For Angelique, that meant going through hundreds of requirements in the RFP one by one.

Working on an RFP is a Herculean task and usually takes hundreds if not thousands of hours for a team to pull together. RFPs sent out by banks that are serious about seeking a new vendor are usually quite extensive, because they are the primary tool to help them decide what kind of company a vendor is and what their track record is. "Continuity is extremely important for a bank and for a regulator. When a bank is looking for a new vendor, they want to know the company won't be insolvent in two years' time. In an ideal world, they want to work with a company with unlimited resources. Basically, when a bank sends out an RFP, they are doing so because they want to narrow down the topics they can be anxious about," Matthijs explains.

The point, of course, is not to "test" the vendor or make them jump through unnecessary hoops; it's so that they can be absolutely sure their customers' data is in good hands. Some RFPs are extremely extensive. They will ask in great detail all about the inner workings of the company. *What's the stability? What's the growth? What are the key contract provisions? What kind of accounting principles does the company operate under?* Ultimately they are assessing the feasibility of the delivery of services, the infrastructure and operating models, and whether or not the vendor can provide a positive and seamless experience from day one all the way through every phase of growth, as well as the vendor's ability to understand their own customers' needs. Most importantly, they will also be assessing the ongoing control and governance (and want proof or evidence) as well as risk mitigation. In addition to proving all of this, the vendors have to provide competitive pricing, expected terms and conditions, workable timelines, the commercial terms, client service agreements, credentials, references, and a track record of success. The RFP will also usually ask about the vendor's ongoing company strategy—where they are headed, their hiring practices, and how many other clients they have (to assess whether they can meet necessary work goals in a timely manner). In most RFPs, the vendor also has to provide detailed information about the proposed migration and implementation protocols. In addition, the RFP will ask specific technical questions about the use of technology and, most importantly, security. As part of the RFP, a bank will ask vendors to create a mapping between their own security policy and that of the banks. "So, as a single question, they want you to make a mapping of maybe a hundred different topics based on their security policy. For example, they will ask how the connectivity is protected by secure socket layers, what type of principles a vendor uses when developing, how security is managed throughout the software development process where data resides, how backups are made, and how long they're stored, and so on," Matthijs says. Needless to say, it is a massive job.

"Everybody that needs to contribute to this document will come together," Matthijs explains, "and we will agree, internally, how we are going to respond to it, because if there's four hundred questions and we have four people working on it (me, Angelique, Chris, and Bas, for example), we would need to divide the work." And the work takes time. According to Matthijs, "We make sure everybody responds in a consistent way. We then calculate at least one hour per question, more likely two." With anywhere from three hundred to four hundred questions in an RFP, the team could spend anywhere from eight hundred to a thousand hours of writing, editing, and meeting to discuss it.

Such a process is costly as well. Matthijs estimates that each RFP costs the company about €50,000 to participate. "The decision whether or not to engage in such a process is a little bit of a paradox," Matthijs points out. "On the one side, if it was very easy to participate, it might cost you less. But if they ask very little, you know they're not very serious and that you can run into a lot of trouble further down the line. So, you want RFPs to be extensive. However, you don't want to spend five hundred hours on an RFP if there's a lot of uncertainty about whether you'll win. So, it's this puzzle that we need to determine every time." In other words, the team has to be strategic about what RFPs they are willing to spend time (and money) on.

Once the questions are all answered and aggregated, there are often several internal reviews—with up to six or seven revisions. "Most companies follow the 'four-eyes principle.' But at Ohpen, we run it past at least six or maybe eight sets of eyes before we submit it. So the first document is being created by the operational people. Then Angelique reviews it, and then I review it," Matthijs says.

For this particular client, Angelique, Matthijs, and Chris worked on the RFP nonstop for several months, fulfilling every requirement and meeting every deadline. "Like Chris and Erik did with Robeco, we had to give demos, show the platform, do workshops, explain how we built it," Angelique recalls. Their hard work eventually paid off. The potential client notified them that Ohpen made it to the final round—there were just two other contenders.

Before each demo, Angelique showed up two hours early—making arrangements with the janitors to let her in so she had ample time to set up her projector, check cables, and make sure all the equipment was ready to go. She didn't miss one detail. Angelique remembers, "We felt it was really important to show them that we would go all the way. So we brought our own projector. We tried to exceed their expectations in how we presented ourselves. We were totally prepped." The "we" she was referring to was the entire team that accompanied her in the demos—including Bas, who presented the IT, and Erik, who did the demos and walkthroughs of the platform. Angelique recalls the pitch with pride. "It was brilliant." She and the team felt like they had a slam dunk. However, one day she received a call from a consultant—also called Chris—who said, "Hey, so sorry to hear that you lost that RFP." Angelique was taken aback. She had meetings scheduled in less than a week's time. Something didn't add up. "We were fully preparing, and we were working like eighty hours a week to make sure this RFP was the best we ever did."

Meanwhile, however, Chris knew who the competition was and heard through the grapevine that Brand New Day (BND), the company now owned by Matthijs's and Chris's old boss from BinckBank, was also participating. They knew that the client had even visited BND's offices. Chris and Matthijs wanted to know more about it. Chris told Angelique to call the client to tell them that Ohpen had heard that they (Ohpen) were just competing for shits and giggles instead of the real deal, because "the client had already chosen BND." During the call, the client disclosed unknown information. Chris then met one of the BND guys at a party and started doing what Chris calls a "Jedi mind trick." Chris told the BND guy that the client said all these things about BND, and the guy started saying it was not true and proceeded to tell Chris their whole pitch.

Angelique, Chris, and Matthijs showed up for their scheduled presentation. The board members were there along with the entire client team, and Chris confronted them. "So what the hell are you doing? We hear that you already made your choice. What are we doing here? Is this a paper exercise?" But they kept on going and delivering everything, giving them what Angelique recalls was "a really good offer, a really good proposal." When all was said and done, BND had been eliminated. It was just Ohpen and one other.

Ultimately, Ohpen wasn't selected. Angelique was curious and asked the procurement officer who managed that project to give her some honest feedback. She wanted to know why they lost. *Was it because they were too bold or too direct? Was there something wrong with their functionality?* In the end it came down to price. "Apparently, our competitor was so worried that we would enter the market and start winning RFPs that they offered the proposition at a ridiculously low price," Angelique remembers. "And I believe that if you pay peanuts, you get monkeys. So we didn't even jump in and counter."

Again Ohpen was going to stick to their values and not devalue their services just to compete. Though they lost this RFP, it still felt like a win to Angelique. "It was the perfect RFP. We were fully aligned. From the Ohpen side, everything—our team, our process, our story, our pitch—it was all perfect." She says that to this day she remembers carrying the heavy crate in with the high-end projector because the client's was so outdated. "I wanted them to see the beauty of the platform that Erik built," she says with a sense of pride. "It was the best pitch we ever did and still is today."

OHPEN SIGNS AEGON—CLIENT NUMBER THREE

As is the Ohpen way, no one dwelled too long on the loss. They moved on, and by early January of 2015, they were in talks with

Aegon. Two years earlier, Chris had delivered a pitch at Zwitserleven. It was there that Chris really hit it off with Maarten Edixhoven, who was the CEO. "One day Maarten told me he would also join Aegon and leave Zwitserleven. Maarten told his friend and former CEO of Zwitserleven, Marco Keim, he should really meet me. So we did. We met and hit it off."

It was during this meeting that Marco asked Chris what he would do if he was the chief of Aegon. "I knew what I would do, and I thought it was a really cool and innovative question," Chris says. "So I told him I would create a direct-to-consumer strategy in all European countries. Kind of an ING Direct but for invest-ments—one platform for all these countries. There are one billion people who are ready for this service in Europe alone. This would be the first of its kind! This was before Revolut or Acorns. Before all these new fintechs were coming in and attacking the status quo. With the technology available today, they could offer really good financial products to everyone in Europe at very low prices. It would be awesome, because instead of all these new fintech companies disrupting the market, it would be an incumbent doing that to their own market. This is what Apple would do, not a big financial firm. So getting this done would be amazing. Kind of like offering what is now only available to the high-net-worth individual to everyone. *Democratization of asset management*, we called it," Chris explains.

Chris knew that with Ohpen's technology, they could build this for (and with) Aegon. "I presented this to the board of Aegon, and they all loved the idea. And that is how we started that project. My idea was an investment product for everyone—an automatic invest-ment plan: just set it up, and get on with your life." Chris also told them how they should do it. "Get it out of Aegon—create a different entity, come up with a different name, use different people—and establish a kind of start-up mentality." He recommended they fund it and own 100 percent of it.

Of course, he knew who could do the executing—Ohpen. For several months, Matthijs, Angelique, and Chris met with board members early in the morning or late at night to accommodate their board's schedule. Chris spent the better part of two months completely dedicated to working on the presentations and the project. In June of 2015, Aegon signed a contract with Ohpen, and the team began engineering and building out the international robo-advisor. Aegon committed an international team to the project from eight different countries, and the Ohpen team worked with them from their Amsterdam HQ for the entire summer.

However, it was like déjà vu. With little warning, nearly two years after initial talks began, Aegon pulled the plug and canceled the project. The entire team was disappointed and surprised. "We'll never really know the reasons why large financial companies start projects and then cancel them, because we are not in the room when they make the actual decisions," Chris explains. While he is quick to acknowledge that sometimes companies have legitimate reasons, such as shifting priorities, new leadership, or internal issues, it's still frustrating. "In a company like ours, where people give it their all, we just really like to know. Everyone worked really hard on the project, and they are proud of their work, so they would like to know what happened." Chris still wonders.

Though Chris and the team may never know why the project was pulled, he harbors no hard feelings. "Our business has advantages and disadvantages. This is one of them. It's part of what we do. I think the people at Aegon are very good and are very committed to embrace change in this new digital world. We worked really well together."

So they signed a contract in December 2016 to use Ohpen's platform for the Netherlands savings and investment operations for their incumbent bank, challenger bank (Knab), and pension provider. Ohpen had their third client.

TURNING DOWN A MAJOR OPPORTUNITY—STICKING TO THEIR CORE PRINCIPLES

In that same year, though Ohpen had suffered some hard knocks, they weren't so desperate to go global that they would compromise their core values. During the busy year of 2015, Ohpen was presented with an amazing opportunity. In April 2015, BNP Paribas asked Ohpen to put together a pitch. Matthijs recalls, "In our pitch, we recommended that instead of managing a mutual fund and selling it through third parties, pension funds, and institutional investors, they should sell it directly to consumers through a website. Direct to consumer (D2C) using technology available today." Then, a couple of months after that pitch, the CEO of the Netherlands branch of BNP Paribas called Ohpen and said that the chief from Paris was in Amsterdam and wanted Michel and Matthijs to join him for an informal lunch. There, Matthijs asked the CEO of the Paris-based group if he could come to Paris in a couple of weeks and, as Matthijs puts it, "show you what we've got." They agreed, and Matthijs headed to their beautiful offices in Paris. Matthijs was awestruck by the grandeur and beauty of the place—noting the stained glass ceilings—and the storied past of the building. While there, they agreed to some workshops with Ohpen, but having learned from the past, Matthijs made it clear that the workshops were not to be free of charge. They would need to be paid.

During the workshops, BNP Paribas was very enthusiastic and, instead of doing it only in the Netherlands, they created a European proposition out of it. In the end, they agreed on the scope of the project. However, they had some issues in the States and with one of their previous vendors, who also operated on the cloud. They found out that Salesforce, contrary to what they'd previously thought, didn't store their data in Europe but rather in the US. As a consequence of that, they basically said to Ohpen, "The cloud is not going to work for us. But we will continue with you, if you are prepared to install your software and deploy new releases in our data centers."

Matthijs recalls, "We were put in this awkward, awful position. We desperately wanted new clients, and they were *a dream client.* But they asked us to betray one of the core principles of the way we wanted to conduct business."

Matthijs and Chris contemplated their next move, and they consulted with Michel. They made their decision: despite how lucrative and wonderful it would be to do business with BNP Paribas, in the end they couldn't in good faith do so. "We couldn't control it anymore. So ultimately we said no." They did so mainly because it was against their own vision and the ultimate purpose of the company—a core banking–engine, cloud-based company being offered as SaaS. As Matthijs says, "In the end, we had to stick to our principles. It wasn't worth it to sign a great client like BNP if it meant we had to give up on one of our core values that underpins the Power of One principle: that

> **In the end, we had to stick to our principles.**

we're all in one infrastructure with one version of the platform." Chris adds, "We were very lucky that our investor stood behind us. These are moments where you can get into a conflict very fast, but we didn't. We all agreed immediately."

CHAPTER 12

GOING GLOBAL— THE FIRST UK CLIENT

WHILE CELEBRATING SOME WINS, the team was also taking stock of the lessons they'd learned from RFPs and the clients they'd lost. By not landing their first UK client, they began to realize, as Angelique says, that "the Brits were never going to sign a non-UK-based company." The Brits had long been the elusive brass ring—just out of Ohpen's grasp. Chris had been in the UK, meeting with potential clients all throughout London, since the inception of the company in 2009. Chris recalls, "We always knew it was the biggest mutual fund country in Europe. It was, together with the Netherlands, the first country that changed the laws so that mutual fund advisors had to be independent. Because we knew this law was coming, change in the industry was inevitable."

After they lost the major UK client at the eleventh hour, Chris immediately began speaking to Michel about scaling up, getting extra funding in, and building out the UK properly. Meanwhile, Angelique and Matthijs concluded that in comparison to other EU countries, Britain was risk averse when working with small outsourced companies. In the spring of 2016, Angelique had an epiphany of sorts and went to Matthijs and said, "We need to show these people that *we are committed*. We have to have an office in the UK *before* we sign clients." When Chris and Matthijs discussed who to send over to the UK to run the office on the ground, Chris knew it had to be "someone who gets shit done—so … Angelique." Meanwhile, Chris would focus on operations, clients, and technology from the Netherlands. And Matthijs, who'd had experience opening an office in a foreign country while at BinckBank, knew what to do to get the process started. He and Angelique spent several evenings making lists of all potential clients and researching offices. Matthijs remembers thinking, "We were going to get an office, regardless if we had meetings or not. We were just going to make sure that we, either she or I, were in the UK three days out of five days. Our thinking was that if we ran into somebody or we heard of something happening at the last minute, we could be there and at least one of us could attend the meeting." While at all those prospective meetings, Angelique and Matthijs heard the common refrain that Chris had heard when he was out on sales calls: "Brilliant idea, but we want to see it live here (meaning in the UK) first." In other words, Angelique and Matthijs needed one key, successful UK client if they were going to expand there. For Angelique, landing one UK client was everything, because if they got one, the potential in the UK would be unlimited. "Every single prospect I met with said to me, 'If you can get this to work, you will blow away the traditional company IFDS, which is paper based. And you will blow away FNZ, because they don't deliver their projects. If you can deliver a digital proposition, and deliver it on time and make it work, it will blow all the competition in the UK away—and you will conquer the UK.'" Matthijs adds, "From the start, we knew the UK was the mother of all countries in Europe, because there's so many investment propositions, so many players, so much wealth, and so many consumers. So it was basically the Valhalla of our business."

And as the story goes, to get to Valhalla, one has to go to battle first. And that's what Angelique and Matthijs geared up for. Angelique packed two suitcases containing hardware and flew from Amsterdam to London to set up the Ohpen UK business.

> **As the story goes, to get to Valhalla, one has to go to battle first.**

Angelique's first step was setting up a job-sharing space, and they rented a two-desk office in a 1.5-square-meter office. From there, she started the hunt for the "real" office as well as for an apartment for herself. She also needed to set up a UK bank account and to start filing all the necessary paperwork to operate in the UK. Her task was simple: set up the company, hire a team, sign the first client, implement it, and achieve a growing client base.

Within two weeks of moving into the shared space, Matthijs met with a business relation who told him Invesco was about to send out an RFP. As if by a stroke of luck, Invesco's office was only two hundred meters away. Matthijs invited James Rawson from Invesco to visit their UK office and view a run-through of the platform, including a demo for the front end (a website for consumers). Matthijs remembers watching the man's jaw drop when he saw the fully integrated functionality. James was also impressed by their shared office space. He saw Ping-Pong tables, pizza boxes, and young people working in an agile environment, and he liked what he saw. Then he said to Matthijs, "It's too bad you don't have live clients in the UK, because that's something we require for the RFP that we're doing. However, I like your solution so much that if you send me a deck that outlines what you're planning on doing in the UK, how you're building the organization, and how you are ensuring regulatory compliance, then I will make sure you get a fair shot of competing for the RFP."

That's all Angelique and Matthijs needed to hear. They spent the next couple of evenings building a deck containing their go-to-market strategy and sent it over immediately. Invesco invited them to complete the RFP, which included over five hundred questions. Chris, Angelique, and Matthijs locked themselves in a room for three weeks and answered all of them. Once the RFP was submitted, Angelique was busy trying to get authorized by the Financial Conduct Authority (FCA). She made it a personal goal to get authorized within six months. Everybody around her in the industry told her, "That's nuts. You can't do that. That's impossible." But Angelique just shook off their negativity and said, "I'm going to try." She had to do more than *try*. She had to get the FCA authorization, because it was required by Invesco as part of their contract. If they were going to land Invesco, they needed the FCA license. She and Matthijs had several meetings with the FCA over the course of several months. During their final meeting and presentation, Angelique looked over at one of the four people with whom their fate was placed, and he was sound asleep. Angelique recalls with a shocked tone, "This was not a 'blink your eyes because you're tired' kind of sleep; this dude was taking a proper nap." Angelique panicked. She went back to the office afterward and spoke with a coworker, Nikhil Sengupta, her first hire in the UK and the one who'd once worked at the FCA, and asked him what it meant if someone fell asleep during an FCA presentation. He assured her it was a *good sign*. "If he doesn't have any questions, he just naps." Angelique couldn't believe it, but Nikhil was right. They soon got word that they were authorized to operate in the UK, but they had to do some final paperwork—which included having £1.2 million in a UK bank account. Seemed easy enough. So Angelique went to their bank in London and requested a verified bank statement that proved Ohpen had £1.2 million. When the bank associate returned the statement, it said £1,999,993.50. Angelique looked down at the certificate and asked why it wasn't for £1.2 million. They apparently had deducted a transaction fee. Angelique felt the queue behind her growing and the bank manager wanting to finish quickly, but she was not going to jeopardize this because of £6.50. So Angelique asked to set up a transfer on her mobile for £6.50 from *her own* bank account and deposited it into the Ohpen account, and then she had her £1.2 million verified by the bank.

"You can still see the deposit made from Angelique Schouten on the account," she adds. "I gave them my two pence, or in this case my six pounds and fifty pence."

It was a personal victory for Angelique. She proved the naysayers wrong and did indeed get the FCA approval in under six months. And on a cold gray day in January, she received the official documentation. Feeling in the mood to celebrate, Angelique announced to her UK team, "We are here to stay! We're authorized." She suddenly felt the spark of an idea and told her staff, "Let's welcome the UK and send a bouquet of tulips to our top prospects and some journalists. Let's make this country a bit more colorful." She bought bouquets of Dutch tulips for £14.50 each, handwrote personal letters, and invited the prospects and press to contact her. She instantly got a message from someone at another UK bank. Others weren't as receptive. In fact, one large bank's CTO emailed her back and said, "We are not allowed to accept anything. Please pick up the flowers within twenty-four hours or we will destroy them." Angelique had a good laugh but emailed him back and explained she was just being kind and to please share the flowers with someone on the way out of the office to brighten their day.

During this time, Invesco visited Ohpen's Amsterdam offices, and Matthijs and Angelique attended many meetings and workshops and were taking the lead. Chris had been saying for years, "I am the only one who got revenue into the company. Every dime we ever made as revenue, I pulled in." In Chris's mind, running an operation is easy compared to sales. "Selling is the most difficult thing," he adds. Chris told everyone many times that to get to the next level, "you have to get revenue in for the company." Matthijs, understandably, was fed up with Chris mentioning this, by Chris's own estimation "one too many times." So when Chris and Matthijs heard about Invesco, the two looked at each other and asked, "Who is going to be in charge?" Matthijs volunteered and said, "I got this." Chris backed off and let him and Angelique get the client, and they did. When all was said and done, Invesco awarded Ohpen the contract in the early summer of 2016. "That was the moment I realized that Matthijs could run the whole thing if I decided to stop. He stepped up to the plate, worked his ass off, and signed the client. *He's got this*," Chris says.

It was a busy but exciting time. Back in Amsterdam, Lydia was getting married, and Chris was about to undergo surgery. Matthijs agreed to spend the week in Reading and go through the contract negotiations on his own. At one point during the week, he received a markup of the agreement. Before sending it over, one of their lawyers said, "I like the contract—it's almost a hundred percent perfect—however, since you guys aren't native English speakers, there are a few minor things we need to change." Chris and Matthijs were happy to hear this, but then they were shocked when they received the actual markup. Chris adds, "It was the biggest markup we had ever seen in our lives. They rewrote the whole thing and wanted to meet the next day to discuss!" All told, the markup consisted of some two-hundred-plus changes—and they had sent it over at five in the afternoon. "These are the moments that separate the men from the boys, and I had no doubt in my mind that Matthijs was up for it," Chris adds.

Knowing that time was of the essence, Matthijs spent the entire night going through all the changes and divided them into categories: approved, not approved, and for discussion. He showed up at the Invesco offices the following morning with the entire document with him. The Invesco team couldn't believe it had all been processed in a single night. The team reiterated several times throughout the project that they couldn't believe what Matthijs had done. After the contract was signed, the entire Invesco legal and compliance departments decided *that day* to go with Ohpen as a supplier. Invesco experienced their first taste of Ohpen's give-it-all and exceed-expectations mentality.

All seemed to be going well, and then on June 23, 2016, Brexit happened. The referendum calling for the UK's proverbial divorce from the EU passed, and markets around the world reacted. In the US, the Dow dropped nine hundred points in two days. The pound dropped to its lowest point in three decades, and European stocks plummeted as well. Large banks in the UK suffered huge losses, and shares in Barclays (BCS) and RBS (RBS) were halted due to extreme selling. In an article for CNN Business, Kit Juckes, a strategist at Société Générale, said, "The extent of the uncertainty that now clouds the UK's economic and political outlook is hard to exaggerate."[7] There was no way to understate it: the future of financial institutions in the UK had changed seemingly overnight. It didn't worry Chris, because you always need an administration whether you are in or out of Europe. Matthijs waited to see if Invesco would react or pull out at the last minute. Their team made a thorough analysis and believed there were no issues. And on July 1, 2016, after just a few short months—from initial meeting through due diligence to contract signing—Ohpen was now in charge of BPO for the seventh-largest asset manager in the world, and they had their first UK client. This was a massive accomplishment, and it didn't go unnoticed by Chris. "Without Matthijs and Angelique it would have never happened. I started thinking," Chris remembers, "that these guys might be able to run the whole thing without me. The first signs were very promising. Landing this client in the UK was the biggest achievement of both of their careers by a length."

A ROYAL INVITATION

Rather than do a traditional contract signing, Invesco invited Ohpen to a ceremony at the Royal Henley Regatta, held in the scenic town of Henley-on-Thames in Oxfordshire, England, on the river Thames. It's quite the posh event, dubbed (by the regatta's own website) as the "highlight of both the summer sporting calendar and social season" in England. Ohpen, the small Amsterdam fintech firm, had been royally and officially invited to compete in the UK—in a big way. For the special event, Alex lent Chris, Matthijs, Michel, and Angelique the Amerborgh plane. Alex could not come, so Edgar, his son, went along for the first time as Alex's representative. They all dressed in their finest apparel (and Angelique in a hat), toasted champagne, watched Olympian-level crew events, and celebrated their new part-nership. "*It was so British*," Matthijs says with a laugh. Though it was exhilarating for the entire team, it felt exceptionally so for both Matthijs and Angelique.

After Henley, word got out quickly. Ohpen had "arrived" in the UK. Angelique and Matthijs began holding so many meetings with prospective clients that they couldn't keep up. Angelique estimated that they were meeting with approximately sixty prospects per quarter while also implementing their first UK client and having over three hundred interviews to find the right people for the UK office. She recalls, "It was like running on empty, and we were just getting started. Adrenaline was the only thing that kept me going." The pos-sibility of all that the UK represented was the only fuel Angelique needed. Though on her own in London, Angelique knew that Chris trusted her, and he assured her constantly that she could handle it. Matthijs was still splitting time between both offices, because he was also needed back in the Netherlands, where the Ohpen team was busier than they had ever been. After all, they had two full clients and were in the middle of implementing Aegon and NN while working on another RFP—this time for de Volksbank.

For the next eighteen months, the team worked to build the BPO platform for Invesco—allocating enormous resources to launch the project. Lydia was tapped to follow in Angelique's footsteps in the UK and to be the head of the London office and help lead the team. After being Chris's right-hand person for seven years, Lydia had Chris's confidence that she was ready to take all the things she had learned from watching him over the years and "add her own swing to it." Additionally, after watching what Angelique could do in the UK, Chris thought she too was ready for the next step. He wanted to put her in charge of all marketing, sales, and PR—essentially taking those responsibilities from him—and he added her to the executive board as well. As far as Chris was concerned, they were both ready for the next level. He knew that if they both succeeded in these new endeavors and served on the board alongside him and Matthijs, eventually they would all be poised to run the company without him.

Client number two, Nationale-Nederlanden.

First team to become
Ohpen shareholders.

What an awesome
experience sponsoring
Mark Slats.

Top: CEOs of Brand New Day, Bux, Blanco together with Matthijs and me. `
all started twenty years ago as the four lieutenants of Kalo and we still see ‹
other every year when we spend a weekend together in Ibiza.
Bottom: We always enjoyed having Ilco (on the left) eat something he ha
never eaten before. This time his first paella.

op: All the important tech decisions were made in Ibiza, where every year
we rented a house in autumn to talk about tech for sixteen hours a day.
Loved these weeks.
ottom: Work hard but also play hard. Eating fish on the beach in Ibiza with
the tech team.

Monkey Money Mind was the book we made together help others to avoid financial mistakes and help supp animals along the way.

Scan the QR Code here to find more photos of Ohpe and the team.

CHAPTER 13

OPENING BARCELONA: BUILDING THE FUTURE OF CLOUD BANKING

IN 2016, WHILE ANGELIQUE AND MATTHIJS were busy building their UK presence, Ohpen was also looking into how to expand their workforce globally. "We always knew that technology would take over the job market and we wouldn't have enough programmers here to go around," Chris says. However, looking for and attracting talent in technology is costly. "Programmers have all these possibilities. American companies are here in Amsterdam—Salesforce, Facebook, Amazon, Google—and they can outbid us on anything. They can pay huge salaries. They can give big, beautiful stock options—and their stock has been going up for years. Then there are all these Dutch companies—Bol.com or booking.com, for example—that we have to compete with for talent."

Chris acknowledges that while Ohpen is a great place to work and on the vanguard in the fintech industry, it takes work to keep up and remain competitive let alone scale in such a competitive hiring environment. A solution they came up with was to look outside Amsterdam to create what Chris calls a "tech hub"—or a development center. Not ones to do things like everyone else, Chris and Matthijs decided to steer clear of Eastern Europe, where most companies built their own hubs because the labor was cheap. "Again," Chris adds, "we weren't going to be different just for the sake of being different." There was a reason behind going south rather than east. In Chris's mind, every major city in Europe had a pool of talent to draw from. "It doesn't have to do with culture. It doesn't have to do with languages. It's a fact—every big city has good programmers." Convenience, accessibility, and proximity became a driving force, not simply the cost of labor. Though in truth, Barcelona was competitive with Eastern European countries in terms of the cost of labor. (In 2016, Spain had a very high unemployment rate—nearly 20 percent—and it has since dropped to 14.5 percent). Not only was Spain a good market for hiring, it was also an easy flight from Amsterdam as well as Ibiza, where Chris has a home. From a lifestyle perspective, Barcelona has always been a great city for professionals. It boasts consistent sunny and warm weather and has a unique vibe. In many ways it is similar to Amsterdam. Barcelona is also multicultural and diverse, attracting people from all over the world to its historic yet modern city. It's also a family-centric and health-conscious city. It is consistently hailed as one of the healthiest cities in Europe—people bike, walk, and make physical activity a part of their daily lives. Healthy eating is a part of the culture as well. Though some in the company were hesitant about moving to Barcelona at first, warning Chris of the vastly different cultures (the Dutch are known for their planning and sticking to schedule and the Spanish for a more laid-back work environment), Chris was undeterred. "We at Ohpen have a very different culture than typical Dutch companies. Besides, I believed we would find people who fit in anywhere," Chris asserts.

With Barcelona settled as the location of choice, now Chris and Matthijs needed someone to take the reins and open the new office. Chris knew exactly who to go to: Jan-Willem. He had the precise mentality Chris was looking for. Chris recalls a time during the first couple of years of Ohpen that stuck with him. Jan-Willem wanted to take exams to get his stockbroker certification and another that would help him understand the exchanges and the financial world's issues. The problem was that the books needed to study for them were enormous, and it would take over a hundred hours to properly prepare for the exams. Between working long days at Ohpen and playing for an elite soccer team five nights a week, Jan-Willem had no time to study. He went to Chris and asked if he could have two weeks off to study. Chris agreed, and Jan-Willem spent every waking hour in the library for two straight weeks, took the exams, and ultimately passed. This kind of focused determination was what Chris was looking for in the person he tapped to lead a new office. "I needed someone who really wanted this and had an enormous drive, someone I could trust but who also could connect with people really well. JW was the guy!" Chris said. Though he knew Jan-Willem was young, he also knew he was up for the challenge, and age has never been a determining factor for Chris, who was also young when he led teams and started his own company. Most important in his decision-making was that Chris had witnessed how Jan-Willem handled the NN project and felt he had what it would take to lead and get a project off the ground.

Much like Angelique was doing in London at the time, when Jan-Willem arrived with his wife and three-month-old son on November 10, 2016, in Barcelona, he, too, was given carte blanche to find a new office, build and hire a new staff, and establish a working Ohpen culture that fit within the Barcelona lifestyle. Since Angelique had moved to London three months prior and in some ways paved the way for branching out in different countries, Jan-Willem asked Angelique for an action list of things she'd used to get the office in the UK opened. However, Jan-Willem quickly learned he would have to make one of his own. "I immediately checked her list against the rules and regulations here in Spain and was told by a friend of mine who had lived here for five years already that I basically had to throw the entire list away." What the entire team was learning was that the rules and regulations for one country don't necessarily apply in other countries. Expansion is more difficult than it seems. Jan-Willem knew then and there he had to "reinvent" himself and the way they would do things in Spain. He recognized very quickly that it was difficult to plan in Spain. Things he thought would only take a few hours or a few days in the Netherlands or the UK ended up taking months in Spain. "Because," he says, "of the bureaucracy." Jan-

Willem—a tall, thin, extremely fit, and health-conscious individual who rises each day at 5:30 a.m. to ride his bike up a nearby mountain and a self-professed "motivated" person who likes to have things under control—quickly realized that his plans weren't working and he didn't have a lot of control over anything. "The first year was the hardest, because I couldn't get things done." Jan-Willem was feeling the pressure. "The board up in the Netherlands was asking, 'Why isn't there a bank account yet? Or when is the company starting?' And all I had was, 'Yeah, I don't know.' Sometimes it was the only answer I could give." He would explain the situation and say that he'd been told something would take a week but, as he says, "You never know in Spain—a week could mean a month later." Another obstacle besides the slow-moving bureaucracy was that Jan-Willem was not a native Spanish speaker, and most people in Barcelona don't speak English. Faced with communication differences, cultural differences, and bureaucratic differences, Jan-Willem's patience was being tested. Fortunately for him, though, the board ultimately understood, had his back, and gave him a lot of freedom to do what he needed to do to get things on track.

FACING CHALLENGES

When he first arrived, Jan-Willem operated out of a rented desk in a coworking space, started to learn the language so he could write Spanish labor agreements, and spent a lot of time on LinkedIn looking for what he calls "the perfect first Ohpen employee in Spain." Once he found that employee, Joan, he had a new challenge: He had to make sure he was fully blending with the engineering team back in the Netherlands. This too proved to be a challenge. "None of us were used to having a satellite office. Sure, we had an office in London, but London wasn't an office that works on a day-today basis with the same procedures, policies, and work instructions with the developers." One of the main issues, Jan-Willem noticed, was participating in the morning stand-ups. "There would be ten people there and two people here, and they would start talking without even noticing we were still there on the screen." Though it seemed insignificant, Jan-Willem didn't underestimate the power of communication and face time with fellow employees for the success of the new endeavor. If this was going to work, they all had to be able to communicate effectively.

Used to being the middleman or single point of contact to communicate the needs of two different parties, Jan-Willem knew what he had to do. By January of 2017, Jan-Willem established some new goals. For one, he would find an office, expand the number of engineers on the team, and—most important on his list—develop a more viable and systematic communication strategy. "They were very busy in the Netherlands, and they needed us. But because they were very busy, they didn't have time to onboard us." It was a catch-22. This was particularly problematic, because Jan-Willem wasn't an engineer himself or a technical person. "It was difficult for me to explain the technical choices we (Ohpen) made in the past, and I couldn't translate them to my new team. Only another developer could do that." This is when Jan-Willem realized he needed to get an experienced Ohpen developer from Amsterdam to help with the new team in Barcelona. This was the turning point. A tech lead moved from Amsterdam to Barcelona, and then shortly afterward a few others moved down as well. Jan-Willem's hope was that the Amsterdam developers would serve as a bridge between the Amsterdam and Barcelona offices. His ultimate goal was that all the new hires would be on the same operating level. He also managed his expectations. He knew they wouldn't be able to jump in and execute a major project; rather, his focus was on making sure the team could deliver smaller projects.

Jan-Willem knew that though there were obstacles and roadblocks, these were simply growing pains and par for the course when a company scales. In many ways, Jan-Willem was very aware that he was writing a manual of sorts for how to do business in new countries. He knew the company's goal was to expand into other countries, and he was taking mental notes about dos and don'ts and pitfalls of expansion. "You won't know what problems you will need to solve unless you have faced the problems already." In some ways, Jan-Willem recognized he had an advantage in Barcelona over what Angelique was doing in London. Angelique was opening an office to service a specific client. The risks were extremely high and she was working on specific timelines driven by the client. However, because he was rolling out a tech hub in Barcelona, Jan-Willem was afforded a level of freedom to take his time, build the right team, and give them time to catch up and adapt to the Ohpen culture. Patience, ample time, and freedom were key if they wanted to succeed.

ASSESSING GOALS WHEN SCALING

By the beginning of 2018, Jan-Willem had new goals. He analyzed, evaluated, and drew some conclusions about what some of the best practices were that he could implement to help scale and grow more efficiently. One of the key indicators for success, he believed, would be communication and alignment with the main office. Making sure the Barcelona team felt fully aligned and part of the Amsterdam team was a main priority. One of the ways Jan-Willem addressed this challenge was by rolling out a travel overview for the year. He brought his idea to the board: "I wanted to have people coming from the Netherlands to Barcelona for a couple of days each month and have people from Barcelona go to Amsterdam as well." The first person to come was Ilco. In true Ohpen fashion, Jan-Willem introduced him to Barcelona with a sports challenge. "We all cycled up the mountain together to get to know each other." There was a bit of strategy in Jan-Willem's thinking. Instead of having a meeting in the office, he wanted to prove to Ilco his team was up to whatever challenges Amsterdam could throw at them. Jan-Willem was well aware of the stereotypes and perceptions the Dutch had about the Spanish. "The perception is that people here in Spain have a good life. It's always sunny, there's a lot of national holidays, and some think that because of this, people here aren't as willing to work hard." In Jan-Willem's mind, the people he hired were very focused and had a great work mentality. In addition to proving their stuff out on the bikes, Jan-Willem made sure the teams had plenty of time back in the office to run demos and ask questions. Building a rapport, increasing communication, and aligning goals ultimately helped Jan-Willem turn a corner. "I found out that really helped. Seeing each other face to face was needed. That was a good thing to do. And I recommend it to anyone who is scaling."

On a day-to-day basis, Jan-Willem focused on keeping the communication lines open. "We follow the same hours, ceremonies, procedures, and work instructions as the team in the Netherlands. We do the morning stand-up with each team here, and then after that stand-up, we have a stand-up with the leads in the Netherlands. That's a joint stand-up across teams. We work the same way. We work also with Scrum (which enables team collaboration) in an agile way. And we present every two weeks."

USING DEVOPS TO SCALE

Another process that is helping Ohpen to scale and helping Jan-Willem's team fully integrate with the Amsterdam team is the application of DevOps. "Basically, all across Ohpen we're working with DevOps, which for us means end-to-end responsibility." The Barcelona office adapted DevOps as a working strategy before the rest of the company just by nature of the setup. "The cool thing about this was," Jan-Willem says, "for the first time, *we* were able to share our experiences with them." The basic idea behind DevOps is that it is a cross-functional means of working with coding, building, testing, releasing, configuring, monitoring, and delivering. While Ohpen was already using an agile approach, which primarily addresses the communication between teams or between the customers and the developers, DevOps addresses the gaps in communication between developers and IT operations. As Jan-Willem explains it, DevOps is a way of working that combines development and operations (hence the name). "Simply put, that means when a question or a change comes in, from the moment the team picks it up then all the way to production—from development, testing, deployment, security, and maintenance—it's all on one team. If something happens with that project or feature that is delivered, it's that DevOps teams' responsibility. They should fix it. And they must feel the pain of delivering it instead of the incidents going to another team." That is where domain responsibility comes in. As a means to scale the company, Ohpen has implemented DevOps across the organization based on domain responsibility. Rather than divide teams by clients, which is not scalable as the company grows and serves more clients, it now has DevOps teams focused on various domains—savings, payments, reporting, investments, and so on. "Here in Barcelona, we have the domain responsibility for reporting. Each team, whether here or Amsterdam, is responsible for a domain."

Chris, who was an early proponent, agrees that DevOps is the future. He says, "The main advantage in DevOps is that *if you build it, you run it.* One hundred percent of the responsibility falls on the team building it. No more 'It is not me, it's him.' Anyone who ever worked in IT knows what I mean by that." Chris realizes the move to DevOps was difficult for some. He admits, "I am a perfectionist, and that is really annoying for some. Especially since I keep changing my organization until it is exactly how I want it. But if I call the shots, then I am changing it until it works. And *It. Just. Works.*"

EXPANSION AND EMBRACING THE OHPEN CULTURE

Now that Jan-Willem had a team, a system, and an effective means of communicating with the Amsterdam teams, one of his main priorities was maintaining the Ohpen culture while assimilating an entirely different national culture into the fabric of the company. It was not a surprise to Jan-Willem that one of the main obstacles his new Barcelona hires had to overcome when dealing with the Dutch was their very direct communication style. "If you're not from there, it can be very hard if you're a new person." Jan-Willem admits that the Dutch culture isn't the only way to operate, and he quickly noticed the best approach was to soften some of the incoming comments. "In the beginning, I took a lot of the pain. I was kind of the shield for a lot of people here. I asked that everything go through me, and I would communicate any requests or commentary to my team. *And* I would do so in my own way, but it would be just as effective." However, he no longer has to do that. The Barcelona team has adapted to the Dutch direct style, and in some ways the team in Amsterdam has learned to better deliver their requests in a more collaborative way. It's a good thing, too, because there is no way Jan-Willem could act as a shield in perpetuity.

In the process of working on the team's communication and helping the Barcelona team adapt to the Ohpen way of doing things, Jan-Willem was keenly aware that the Ohpen culture—giving it all, exceeding expectations, having fun, and enjoying the ride—could remain even if the style and delivery were a bit different. "I grew up with both the successes and failures of Ohpen. I've been here for ten years. I know our procedures. I know our clients. I know the leaders. Of course, the basic values are very important to me. But there are other things that I want to do my way, and I have been given the freedom to do that. That is very Ohpen. That is actually fundamental to the Ohpen culture—it embraces a variety of personality styles and approaches—as long as the results are the same."

One of the ways Jan-Willem is adapting this particular part of the Ohpen culture is giving people a lot more responsibility—which he enjoyed when he started at Ohpen. "I see my role as more of a manager meant to support them and to create the best environment for them to be able to do their work in the best way, and also to make sure they have all the information they need. And if they can do the best they can, then I've done my job well." Like for all the leaders in the company—they share the same core drive and give-it-all mentality—within that is a lot of freedom to be the kind of leader they are best at. Angelique, Lydia, Chris, and Matthijs each have unique personalities and different approaches when it comes to leading others, and Jan-Willem is no different. "For me, communication is the most important thing. It's right up there next to the core values of Ohpen.

> For me, communication is the most important thing. It's right up there next to the core values of Ohpen.

Whenever I am onboarding new employees, I tell them that next to the core values of Ohpen, to *me* the most important thing is honest communication. *Honesty. Communication. Having fun.* Those are the three most important things for me. And, of course, they overlap a lot." According to Jan-Willem, "If you don't enjoy the ride here, then you'll never succeed. It's very important."

Another reason Jan-Willem has to hit home the importance of direct, honest communication in Spain is largely because of the cultural differences. While the Dutch may be too direct, he notices the Spanish could use a little help being more so. He recognizes that typical Spanish companies are much more hierarchical. Spanish employees are not used to telling managers what they think—or even if they have messed up—out of fear of being reprimanded. "The whole purpose of DevOps is to remove communication barriers, break down hierarchies, and empower individuals," Jan-Willem says. "But it's also what I believe in, it's the way I work, and it very much fits with my own beliefs. It also helps create a relaxed atmosphere. It's not that we're not working hard; it's just that everyone is Zen and therefore working in optimal conditions."

LAUNCHING A NEW OFFICE

Finding a permanent office in Barcelona proved to be one of Jan-Willem's more frustrating endeavors. He had a team; now he needed the right space. However, things move notoriously slow in Spain, and Jan-Willem was running out of patience. Chris decided to fly in for forty-eight hours to help Jan-Willem move things along. "Whenever Chris came to visit, good things happened. Because he lives in Spain, he understands the culture and has more patience than others. Chris makes decisions fast too. And when he came, things just happened. All I had to do was show him the problems, and within two days, things changed for the better."

Within the first three hours of Chris's arrival, Chris had Jan-Willem arrange a meeting with a real estate agent who took the two men to fifteen different offices all over Barcelona. Nothing was jumping out at them or felt right. But Chris didn't lose hope. Finally, at the end of the day the real estate agent in Barcelona took them up an elevator to an incredibly unique space—quite literally in the shape of a circle that had incredible views of the city. Jan-Willem said they both knew instinctively the space was amazing and different—just as the Amsterdam Rokin office was. It was perfect.

"We looked at each other and asked, 'Do we feel it? Is this the office?'"

And Chris's response was quick and decisive: "Yes, let's go for it. Arrange it now. We want all the paperwork done today."

Jan-Willem laughs at the memory. "Of course, they weren't used to this in Barcelona, so it was really funny to see the real estate agent's reaction."

Once the real estate agent left to do the necessary paperwork, they called every twenty-five minutes just to make sure it was moving along and would get done that day. By the end of the day, Chris and Jan-Willem had closed on the office in Barcelona.

Jan-Willem was amazed by how efficient and quick the process went, especially after months of trying to do so on his own. "Why I love to work with Chris is that things happen. We make a decision. It can be a good one; it can be a bad one. But once we make the decision, we go for it. We put all our blood, sweat, and tears into it and just make it work. And if it doesn't work, if it's a bad decision, it doesn't matter. At least we gave it our all."

It turned out to be a great decision. Like the Ohpen office in Amsterdam, the Barcelona office used the same interior designer to assist in creating a Zen-like atmosphere. There are now statues of Buddha, a lush Zen garden in the reception area, a working kitchen, a shower for those who enjoy midday workouts (they go to yoga together and even do HIIT on the beach after work on occasion), several spacious open meeting rooms, and even a large cushioned window ledge with pillows for meeting, resting, or if someone wishes, meditating. It is so different from the slick and sometimes soulless designs of most tech companies. Again, not just for the sake of being different. The entire vibe is meant to inspire those who visit as well as those who work there. Like the Amsterdam office, which is in the heart of the city, the Barcelona office is located on the Passeig de Gràcia, right on the southeast chamfered corner of the intersection of the world-famous Avinguda Diagonal. From the second floor office, Jan-Willem's spacious office, with its classic Barcelona bay windows jutting over the expansive sidewalk below, overlooks the octagon-shaped central intersection. He can see his apartment building from his window. And in under twenty minutes, employees can walk to Gaudí's architectural masterpiece the Sagrada Familia to the east or walk south to the historic Gothic quarter and, if they desire to do so afterward, for a stroll by the Mediterranean just a few blocks away from there.

"The reason it feels Zen here," Jan-Willem demurs, "isn't because of me; it's because Chris fueled it. His positivity fueled it. He was positive toward the people I hired and the culture I was creating here within the established Ohpen culture. He understands that people in Spain don't care about money in the same way the Dutch do. Rather, he says, they care about a positive work environment and a good employer. Sure, money is important, but it's not the most important thing," Jan-Willem explains. "People in Spain don't buy a couch every couple of years. They'll sit on the same couch for thirty years, because they care more about being able to buy a cava after work and sitting outside with friends or having lunch with their family on the weekend. Chris sees that, and he feels the positivity in it."

Over the next several months, Chris traveled again to Barcelona several times and helped Jan-Willem problem solve and keep the momentum up. However, once the space was designed and ready, Jan-Willem had new challenges—getting everyone on the team to speak English. Chris hired an English teacher and arranged to have everyone in the office take English lessons together three times a week. Problem solved.

"When Chris was in Barcelona," Jan-Willem says, "it was always very exciting." The two solved a lot of problems while cycling up the nearby mountain, walking through the city center, and even eating delicious Spanish dinners. Talking over all the business opportunities and potential in Barcelona, Jan-Willem says, "made me really enthusiastic."

It also made him feel truly supported. Having the leadership's support back in the Netherlands, Jan-Willem believes, is crucial to the success of Ohpen's expansion. And he believes he has it. Both Chris and Matthijs want to scale, and the Barca tech hub is the way to go. Jan-Willem has shifted his goals—with the office up and running, the Ohpen culture intact, and communication improved, he thinks the Barcelona office is poised to take on multiple domains. "I think the next step, really, is to have *full* end-to-end responsibility here in Barcelona, but this means having the mandate to decide things and the authorization to get things from start to end and maintain it there." Ultimately, the goal is to fill the entire space with DevOps teams.

Measuring one's success at scaling and growing a company is by its nature a bit of a mercurial process. The goalpost for success is always moving. Just when Jan-Willem feels he has met a goal, a new one appears. He also feels a bit like an underdog, but he admits, "I always loved that role. Because from the get-go, you can exceed expectations." Like his fellow Ohpen employees, he likes working from a place with something to prove—to himself, to others, and to his team. But he thinks he's in a great position, and in fact, the entire team is.

CHAPTER 14

FOUR BANKS IN FOUR YEARS

WITH THE EXPANSION happening in Barcelona and the UK office up and running, led by Angelique and Lydia, back in Amsterdam Matthijs and Chris were looking to grow in the Netherlands as well.

In the early years of Ohpen's existence, Chris had been the one in the lead in sales. But by the end of 2015, when de Volksbank announced it was going to migrate their investment accounts to a new BPO provider, Matthijs approached Chris and said, "I'm going to bring in Volksbank." Matthijs did so for two reasons. For one, he knew he was the right fit for de Volksbank. De Volksbank had been nationalized a couple of years earlier. In the middle of the ongoing financial crisis, there had been significant issues within the group with property financing. Eventually, the Dutch national government expropriated the bank's bondholders and shareholders. Therefore, they weren't a bank with a penchant for risk taking or operating with a start-up mentality. In other words, they were very thorough. Matthijs admits, "In Holland, we have an expression that says '*Act normal; that's strange enough.*'" By his own admission, Chris is an energetic and enthusiastic person. As Matthijs says, "He's more visible and present in the room, and sometimes the room isn't quite big enough for him. And I think within de Volksbank, they value a more reserved, measured, and calculated personality." Chris didn't disagree. For Chris it wasn't about egos and who was in charge. In his mind, he wanted what was best for the company, and he welcomed someone else helping him bring in clients. The second reason was strategic—on both of their parts. Both Matthijs and Chris always had an eye on scaling the company. If Chris was the only one taking the lead in sales, the company wouldn't be able to grow at the rate it potentially could. For both these reasons, the two quickly agreed that Matthijs would be in the lead.

It was not an easy or quick deal. In fact, from the start, Ohpen was up against a lot of vendors. "That is a hell of a lot of competition," Matthijs says. In the end, de Volksbank only invited a small number of companies to present and to demo their services to de Volksbank people. After that, every month, they structurally led the process to a decision.

By May 2016, which was six months after the first pitch, Matthijs knew Ohpen had been selected. He optimistically thought, "With just two months of contracting, we will have a finished agreement before the summer holiday." But nothing like that happened. "We were selected, but still there were many things they wanted to know, and they wanted to take many more deep dives," Matthijs recalls. Matthijs took them at their word, but he still had no real certainty. Meanwhile, Matthijs was getting more and more heat from Chris, who was pushing Matthijs to get de Volksbank to sign. So Matthijs went back to de Volksbank and tried to push them, to no avail. They were in no rush.

In the end, they didn't budge. Matthijs wasn't really nervous, because he knew that de Volksbank had received a letter of termination from their existing provider saying that before 2018 started, they needed to be out of the system. "I knew that staying with their current vendor wasn't an option. Still, you can never know, so I wanted the deal to be done." Meanwhile, it was an extremely busy and stressful period—at this time, they were also working on the Invesco implementation in the UK. Matthijs was getting up at four in the morning to fly to the UK several days each week. There were also migrations happening. "We had also just migrated the first investment accounts in early 2016 and were working on the savings implementation for NN."

Then, finally, just before the summer holidays, Matthijs and de Volksbank sat around a table at their offices in Utrecht and formally started the contracting process. They insisted the agreement would be their standard template instead of Ohpen's contract, which was written in Dutch rather than in English (like all of Ohpen's standard agreements). Matthijs was torn and contemplated whether to sign, because he says, "I knew their standard template for outsourcing wouldn't fit our service model and our way of working." Matthijs is quick to pass along to anyone engaging in contract negotiations the following pro tip: "When you have a document that best describes the service you're going to provide to the client, *do not agree to* or do not be tempted to agree to use something other than what you have created yourself. Mapping between someone else's document and your own and making sure things are included that you want there and match what you want to agree with the client becomes the labor of monks. It is a very inefficient process." Matthijs knew he and the team would have to rewrite a lot. In the end, he and Ohpen's legal counsel along with de Volksbank's counsel indeed rewrote the entire thing over the next several months. "We spent three months working on it side by side. We had put this whiteboard in the room where I had the two parties, because I had to supervise both of them, and all the individual clauses. I had a status report that I modified daily to know where we stood, because it was quite difficult to do the two negotiations at the same time and make sure we kept the ball rolling."

In the end, Matthijs told his counterpart at de Volksbank, "If you want to migrate before January 1, 2018, we need to start a project on January 1, 2017. If we want to start on January 1—and we're not going to start if there's no signed agreement—we need to sign the agreement before December 10." In order to meet this deadline, Matthijs and the team had to pull all-nighters. This also meant he had to be in Utrecht by seven in the morning each day. Matthijs admits it wasn't exactly the most efficient way to close a deal. And Chris tried to make light of it and tease Matthijs: "This is the lengthiest contract negotiation in the history of core banking software!" Matthijs didn't disagree, but he does, in hindsight, think that being thorough helped the migration go smoothly—and the entire relationship thereafter as well. When all was said and done, Ohpen signed with de Volksbank on December 8, 2016—almost fourteen months after the process started. "One of the positives of having a long time to prepare the collaboration was that from January 1, we hit the ground running, and the project went very smoothly. We migrated on the weekend of November 14 and 15, 2017—on schedule and without any incidents. It went really, really well," Matthijs says. "In Holland, we say, '*Good preparation is half the end result*,' and I fully subscribe to that expression, because it's true."

IMPLEMENTING DE VOLKSBANK

After a rough start with a newly hired project manager who wasn't up to the task and who was eventually let go, Matthijs and Chris pulled in fellow cofounder Erik Drijkoningen, the now product chief, to help with the migration. Erik was the go-to guy when projects started to derail. When Erik came in on the project, he quickly assessed the situation and noticed that no one on the implementation team had actually written down how the top ten core processes of the bank would be working or how the systems were going to react and interact with each other. "Frankly, it was kind of disturbing to me, because we have all these payments flowing through both systems. All the transactions through each system. And they needed to be in sync for certain things. And no one from either team could tell me at that point in time how that would work. Not our people, not their people. No one actually knew," Erik recalls incredulously.

Erik had to pull a tough-love parenting move and sit everyone down in one room and give them a "talking-to." Then they put Erik in touch with the lead architect of their systems. For the next three or four days, he and the team started drawing and making flowcharts of all these core processes and how the systems would interact together and how money would flow through the entire ecosystem and where reconciliations should be done. "And from then on, everything began working in a convergent way and went pretty well afterward." For Erik, it was all about experience and intuition. "I think it takes a lot of experience or mileage to see where these projects and processes can go wrong. I just had a feeling as soon as I walked in the room and met the teams that something was off."

Once Erik got everyone aligned, he worked on relationship building—also intuitively. After the step-in meeting where he essentially reamed out everyone on both teams, the de Volksbank project leader approached him and asked him how he thought the meeting went. Erik knew honesty was going to be the best policy. "It was a total shitty meeting, but it was necessary to get things going again. I'm disappointed. Because I try to learn something out of every meeting, and I haven't learned anything today." Erik was frustrated and a bit worried about what his counterpart at de Volksbank was going to say in response. But he surprised Erik by stating, "Oh maybe I can teach you something now. What do you want to learn?" Erik, no longer thinking about work, said, "Well, funny you should ask me this, because I was in the shower before I came here, and I live in the woods. And I heard a woodpecker. And I thought to myself in the shower, 'How can a woodpecker make a hole in a tree while hanging upside down?'" Erik looked at the guy and became embarrassed. "He was probably thinking, 'What the shit are you talking about?'" Erik says, laughing. But he said instead, "Today's your lucky day. I've studied biology, and the woodpecker is the only bird that has two claws in front and two in the back, so yes, it can hang upside down in the tree." Erik laughed, and so did his counterpart. That light conversation immediately broke the ice between the two men. From that point on, the two hit it off. "The moral of the story is," Erik advises, "that you need to do something out of the ordinary to get things going." Erik is aware that people sometimes see him as the "joker" in meetings. But for him, it's only to provoke a reaction and to break the ice. "If you can't connect on a professional level, you need to find other ways to connect."

Matthijs wholeheartedly agrees with Erik's approach and credits him with the successful implementation. "Erik is a very likable guy, and a respected guy, and I haven't been in any implementation where the guys on the other side didn't immediately value Erik for what he did, what he knew, and how he interacted with them. What Erik also is very good at is making sure he takes the lead and that all the things he needs to get things started are taken care of. He is a real leader in that respect."

One of the keys to Ohpen's success with clients, Matthijs believes, is the relationships he and the other founders are so great at establishing and maintaining. "Erik is right. Finding common interests—whether relevant to the project or not—is key. Secondly, injecting conversations with some humor can make sure everybody is relaxed and laughs a little bit. Humor defuses tough situations and then helps clients turn their attention back to the issue at hand. I really believe that humor can create hormones in your body that make you more productive and more constructive," Matthijs says. From Matthi-

> **One of the keys to Ohpen's success with clients, Matthijs believes, is the relationships he and the other founders are so great at establishing and maintaining.**

js's vantage point, Erik is pretty good at finding something in common with his clients and making sure everyone is having fun—even when they're tackling weighty projects.

In addition to humor, Matthijs is a proponent of showing vulnerability to clients. He knows that this is a tough concept for some people to wrap their minds around. "I know people could disagree with this. I know some people say you never should apologize, but my sincere belief is that if you look somebody in the eye and say, 'I really messed this up, but my intentions are to make sure that the project doesn't suffer from what I or what my team did,' this goes a long way. Not walking around the elephant in the room but addressing it and saying that you're sincerely sorry can lead to good outcomes." Matthijs learned this firsthand when he was working at BinckBank. At one point, BinckBank suffered losses due to outages, and there was a question about liability and recovery of losses. "The simple fact of sincerely apologizing for what happened takes some of the pain away immediately," Matthijs advises. "And then, once that is covered, you can move on with business."

Matthijs and the team are always honest and quick to admit when they have messed up and are also quick at making amends to get the project on track. When looking back at the implementation phase of de Volksbank, Matthijs is proud of it overall. "I think it's the most successful implementation so far. By objective metrics, having such a long lead time for the project helped. I think the match between what they wanted and what we could offer was a hundred percent. Sure, we did have some tough nuts to crack during the implementation, but I have found that the leadership of de Volksbank are straight shooters."

WISH LIST FOR BANKS WORKING WITH OHPEN

For Matthijs, working with de Volksbank serves as a great example of how banks ideally *should* work with vendors. If he had a list of what he wished banks would know before engaging with Ohpen, it would be extensive. One of the key things Matthijs advises banks to consider is to make sure they properly resource the project—not just with money but by allocating qualified staff that are not, as he says, "running eight projects simultaneously." Having someone's focused attention on such an extensive project is key to the success. They also have to allocate time—not just on the project but in time spent meeting with and working with the Ohpen team. Making sure the division of tasks and responsibilities is very clear is also important. Matthijs also recommends empowering the people in charge to make decisions—and once those decisions are made, they should stick to them. One of the biggest issues Matthijs and the team often face during implementation is that somebody without the proper authority makes a decision and communicates it, and then later someone else disagrees and escalates the situation, and then a couple of weeks or a couple of months later (after an enormous amount of work has been done), the decision is reversed. "It's all about the mandate of the team but also about who does what within the team," Matthijs says. "Within de Volksbank—which is also a pretty complex organization, at least compared to us—they have four different banking labels that all operate in the Dutch market. What de Volksbank did well is they said, 'When it comes to savings and investments, ASN Bank is in charge, so they make decisions with regard to the proposition. SNS Bank is responsible for the IT part.' Both management teams of both banks that are all under the umbrella of de Volksbank agreed that that was the distribution. They accepted that, and they all took to their role very well. It's all about proper resourcing and clear mandates."

Another main thing that Matthijs thinks contributes to the success or failure of a project is that a bank must make sure they select the right tool or the right service for what they want to do. He suggests spending a lot of time determining whether the service providers can do what the bank wants them to do. And after they've selected the service provider, don't ask them to bend over backward to make things happen *exactly* the way you want them to run. For example, during the de Volksbank implementation, de Volksbank accepted what Ohpen proposed in terms of functionality and technicality. And when they suggested things that Ohpen could apply for all of their other clients and make the Ohpen platform better, Ohpen accepted their suggestions. "The principle is that you're not trying to convert a machine to do something it's not supposed to do, or you pull it out of shape. I think any party that is going to outsource certain things should realize that when they want to change the tool they want to use too much, in the end it will be to everybody's detriment, because it will not last in the long run," Matthijs advises.

Chris adds to Matthijs's recommendations and advises, "Don't change the software to fit your processes; change your processes to fit the software. That is so much easier. Second, don't change the scope. You can always change things in phase two. Third, pick a really good, balanced, and experienced team, and decide what the mandate of the team is. *What can they decide? What has to be decided higher up the chain of command?* If you follow these three recommendations, your project will go a lot better than if you don't do this. This is not an opinion. It is a fact. I think we are pretty good at persuading a client to do so." And for the most part, Chris says, clients comply, because they see how much they benefit.

TKP AND KNAB

Aegon, the multinational life insurance company they signed in 2016, also offers pensions through TKP. Aegon also operates a challenger bank under the brand name Knab in the Netherlands. In 2018, Aegon migrated all their saving accounts onto the Ohpen platform, and during this time, one of Matthijs's and Chris's main strategic goals was to move into the pension market. Building out a platform to support pensions became a main priority—and TKP was the perfect client to work with to do just that. In 2017, Ohpen was officially given the green light to migrate Knab's investment accounts. However, around the same time, a major regulatory change was happening. Markets in Financial Instruments Directive II (MiFIDII) was going to affect how Ohpen was going to run their processes. In layman's terms, MiFID was originally set up to ensure that banks were more transparent. MiFIDII was finally implemented in January 2018 to amend this issue. This meant, however, that Ohpen had to adjust their software to comply with the new regulations. Angelique recalls, "It was one of the first times we had to implement such a big piece of legislation and have all clients agree on the rules—all while rolling out an ongoing migration." Matthijs was in charge of making sure the project went smoothly, and ultimately it did. Knab went live on schedule.

A MAJOR WIN: LEASEPLAN BANK

Just after de Volksbank went live in late 2017, Angelique received an email from LeasePlan. They were about to launch an RFP and asked if Ohpen wanted to participate. They requested a meeting, and Angelique invited them to visit the office. It was during their office visit that she determined they were serious about doing business and changing from Oracle. Angelique also knew the stakes were high;

if she won this contract, it meant that Ohpen would have taken on Oracle—*the* Oracle. Not only that, she knew it would be a major win for the entire company, because LeasePlan Bank is a pure savings-only bank with customers in two countries—and it was a very important strategic client to sign. Chris says, "We did not have a pure internet savings bank, and I thought it was very important to have one as a client, because not only is our system perfect for them, once we had one, the rest would follow, and we would get our natural market share."

The original scope was just for the core administration. Then, during the RFP process, LeasePlan became more and more enthusiastic about Ohpen and their capabilities. Angelique worked on the proposal to perfect it. Angelique says, "We showed them other modules and our CRM application within the platform, and they got excited about that."

Nevertheless, Angelique kept her cool. She knew she was in the underdog position. "It would have been the easy choice for LeasePlan Bank to go with their current vendors, Oracle and Five Degrees, while hosting it at Schuberg Philis. In the end, however, LeasePlan liked what Ohpen presented so much that they expanded the scope. "They took more services than originally anticipated, because in the original pitch and scope, they thought Five Degrees would still be a supplier." In the end, LeasePlan completely and ultimately chose Ohpen. Angelique believes the reason they ultimately chose Ohpen was that they wanted something that above all was stable and that worked and secondly that was flexible and reliable. They also loved the Ohpen team. "Where other vendors invited them for yet another deep dive on the solution, we invited the project team for a kickboxing session in our gym. We really connected with them. Even the contract negotiations were super smooth. We hardly had any tension," Angelique recalls.

Things went so smoothly, in fact, that LeasePlan had a poster created to commemorate their working relationship. As a backstory, each Ohpen client is given a code project name during the implementation phase. And for this project, Angelique came up with the name Project DeLorean, from the movie *Back to the Future*. LeasePlan created a fun movie poster complete with a "cast list" of all the names of the people involved. The poster now hangs on the ground floor of Ohpen's Rokin office building. For Angelique, it's a symbol of a project well done. "The poster is a direct reflection of how amazing the project and the relationships were. It was amazing. Everyone on it was amazing." She credits the success of the project largely to the personal connections made during the implementation by colleagues like Kees. Like Matthijs, Angelique doesn't underestimate the power of deep relationship building with clients. And she values listening as well, adding, "There were also no politics involved. Everything could be explained from both sides. There was no tension on who was right or wrong. It was really about explaining each other's viewpoint, and we took the time for that." Like Matthijs, Angelique remained patient during the negotiations and workshops. "On this project, we took the time to explain and to listen to each other. I think that was done very, very well."

Angelique isn't just guessing about LeasePlan's satisfaction with Ohpen. In fact, they've been quite vocal. One day Angelique was on LinkedIn and was reading an article in which LeasePlan was talking favorably about Ohpen: "In Ohpen we have found a partner that allows us to further increase our efficiency and enhance our capabilities to keep track of the market developments and changing customer demand. This will help us deliver on our commitment to bring user-friendly, clear, and competitive savings products to our customers,"

Danny te Brinke, the director of LeasePlan Bank and strategic finance said. For Angelique, this was all the confirmation she needed. "We have a client talking in the press positively about us and who is proud of our partnership, externally as well, without even asking us. That has never happened before!" Then, shortly afterward, Angelique had a meeting with a prospective client and invited LeasePlan to be there. "LeasePlan was basically doing all my work and selling our service for me, saying things like, 'We're so excited to work with Ohpen!'"

All told, by the end of 2017 and in early 2018, while they had some setbacks along the way, they successfully migrated not one, not two, but four new clients, and they proved to their clients, their investors, and even themselves that they really could do more with less.

PART SIX

RIPPING CURRENTS (2018—2019)

CHAPTER 15

FROM VALHALLA TO HELL AND BACK

ALL WAS GOING WELL IN THE UK, but in October of 2018, at the eleventh hour, just as the project was about to go live, Chris and Matthijs were called to the Invesco offices in London for a meeting. They thought it was a follow-up to the governance meetings they had been in during the weeks prior, where they were granted the go-ahead to go live. Chris and Matthijs believed they would fly in on Tuesday and have the meeting, and Invesco, their first UK client, would go live on Wednesday.

When Chris and Matthijs arrived at Invesco, they were greeted in the room by James Rawson, who simply said, "We can only have this conversation 'without prejudice,' so if you want to have this conversation without prejudice, we can have it; otherwise, I'll ask you to leave." Chris was confused. He responded, "I'm sorry, I don't know what that means." Again, James repeated himself. Again, Chris responded, "*What are you talking about?* We've been working for over two years together with you guys. *What does it mean?*"

Another person on the Invesco team spoke up and explained to Chris that as long as he wasn't going to use the conversation in court, they could talk. Chris was dumbfounded. His team in the UK was expecting to go live the following morning. James from Invesco slid a document over to Chris and Matthijs, and Chris couldn't believe what he was reading. They were canceling the contract and didn't want to pay for the remainder of the contract. It was like déjà vu all over again with client M&G, only they'd done it before Ohpen started; this was after two and a half years of work and a €10 million investment. Chris and Matthijs got up from the table and left immediately. Overcome with shock, the two walked out to a nearby park. Matthijs literally felt sick to his stomach. Two years of nonstop work had taken its physical and emotional toll. There had been no real reason given, and it felt like an absolute blow. Matthijs recalls, "We were at the finish line—I could reach out and touch it. We went from complete ecstasy and a sense of invincibility to utter devastation. We were overseeing three parallel implementations—a twenty-five-thousand-hour project, a ten-thousand-hour project, and a thirty-thousand-hour project—working our arses off the entire time and going through dire straits, and that made it so hard for me to accept. It felt like it all had been—at least in the UK—in vain and for nothing."

Chris didn't share that feeling. In fact, he knew the work Matthijs, Angelique, and Lydia had done was nothing short of remarkable. It had put Ohpen's name on the map in the UK. Chris just wanted to get out of there as fast as possible. However, they had an hour's cab ride ahead of them and another flight back. Out in the park, Chris, seeing his friend in pain and literally getting sick from the years of stress, put his arms around him and said, "No worries, brother. We will be fine. Come, let's go home." Exhausted, the two men took the long cab ride to the airport, only to be met with a flight delay. They arrived home in the Netherlands in the middle of the night and went straight to their respective homes. They were so tired.

But before they slept, Chris and Matthijs went into execution mode. Immediately after the meeting, they'd called Lydia, who was running the office, and let her know what had happened. Lydia recalls getting the call from Chris: "I was completely stunned, because I expected them to be in for a two-hour meeting. And they were calling me a half hour after the meeting started. It felt like we were being pulled back and shot with a catapult." Like Matthijs, she felt the physical stress immediately too.

On that very same day, a Wednesday morning, Chris made a plan for the future and told Matthijs what they needed to do. Matthijs agreed but still was a bit in shock. On Thursday, Chris and Matthijs talked to the lawyers to find out what their options were, because they needed to comply with England's employment laws. On Friday, Chris, Angelique, and Matthijs got in a car and drove to Belgium to meet with Alex and Michel and told them their plans: they were going to close the UK office. The employees they had there would no longer be of use without Invesco. On Saturday and Sunday, they got things ready and decided that by the end of the year, they would have executed the whole plan.

Chris knew this would be difficult for Angelique and Lydia, because both had given so much to the UK office. Lydia had had her first child in London. Her life was there now. She also loved her employees and what she had built there. It seemed like an impossible task to have to let go of the people she had grown so close to, and she herself had to move back to Amsterdam with so little warning. Lydia pleaded with Chris to slow down, but Chris was resolute and said to them, "I understand you're very emotional, but we have to be rational and businesslike. We have to fix it." While Chris empathized with them, he knew it was his job to be "the leader" and stay, in his own words, "focused, sharp, calm." He knew remaining calm would allow him to see the future of the company more clearly. And what he saw was opportunity rather than disaster. He felt strongly that there were still so many opportunities in the Netherlands, especially in the pension market. Chris's eye was still on Tier 1 banks, and he knew the moment they got a Tier 1 bank, Ohpen would reach a completely new echelon. All in all, it was not entirely a bad thing to redirect course, close down the UK, and focus on Tier 1 banks and the home market. The international strategy was still alive, but they had to focus on the home market for the next year.

Lydia hung up the phone with Chris in shock. Looking back on that day, she says, "I don't think I really understood at the time what was being said." She turned to her fellow board member, Ian, who was working with her in the UK, and the two went into execution mode as well. They gathered the team together, updated them, and then asked them to go home for the rest of the week until she knew what Chris's and Matthijs's definitive plans were. Over the weekend, she, Matthijs, and the team worked day and night to get everything ready from a legal and HR perspective. On Monday she texted the staff to come to the office. Most of them could see the writing on the wall and knew what to expect when they arrived. To Lydia's and Matthijs's surprise, everyone was completely understanding. They gave her and Matthijs hugs and handshakes, and afterward, in true London fashion, they went to the pub to have one last drink together. Lydia could hardly believe it. She recalls, "People were thanking *us* for the opportunity and telling *us* they were sorry." She also felt a sense of teamwork that was quintessentially Ohpen—proof they had brought the Ohpen's give-it-all mentality with them to the UK. She had employees say, "We did everything we had in our power to make it work, to make it happen, to be prepared for going live." They felt the disappointment *as a team*. In fact, though she felt so badly for them, it was she that received encouragement and support. "People were texting me weeks afterward, asking me how *I* was feeling. That showed me how much people cared. And in such a dark time, it was a positive, small light."

MAKING SENSE OF THE UNEXPECTED

For the second time, the Ohpen team had been bitten by a UK client, and they weren't going to jump back into those waters anytime soon. But it was hard to parse out what lessons there were to learn. Everyone was truly stunned by the sudden about-face from such a once-happy client—one that had celebrated them with champagne at the "social event of the year" and had been meeting with the team for the better part of two years. Everyone has their own theories, but no one believes it was because of negligence. Chris thinks he "underestimated the culture of the Brits" in that Amsterdam tends to foster very direct and honest environments, and in the UK, people tend to be polite but may not be expressing their true feelings or talking about crucial issues. Chris adds, "It was not an environment that I felt at ease in. I am a very pure person, and I only have one life to live. With Invesco leaving, I was able to focus back on positive energy."

In Chris's mind, there is no point in rehashing it. He'd rather leave that to the lawyers. Angelique, however, felt it had a lot to do with laws that were happening in the UK at the time. When she and Matthijs entered the UK market, they did so because they thought the D2C asset management market was going to be revolutionized, because the FCA was going to force fund managers to reduce their costs and fees and, in order to do so, digitize their legacy paper-based channel. The fund managers started D2C projects the minute the FCA announced they were starting these reviews. Hence Invesco's RFP. But the first study was more lenient and not as forceful as expected, and the second one in 2018 for investment platforms was also not that strong. So when the FCA stopped pushing investment managers to digitize and improve their offerings to retail investors at lower costs, fund managers and investment platforms felt no need to comply and began pulling their D2C projects. In other words, Invesco most likely didn't pull the project because of something Ohpen did or didn't do; rather, they *didn't have* to go through the transition. Ohpen wasn't the only one that was affected. Angelique contends, "Most D2C projects of asset managers at our competitors stopped in the UK. All these initiatives stopped, because regulations didn't force them to lower their costs." For Angelique, the UK project, though extremely painful, wasn't entirely in vain, however. "Personally, I've learned so much. One, I learned about myself, that I can do anything in any situation—I could build the company overseas. I will make it work with what I have, which is a strength. And like most people here, they all made sacrifices. They believed in this company. It was fun, you know? We had really good times. We enjoyed it. Bonding with colleagues in the way we do isn't normal at most companies. We take it for granted sometimes. But because of Invesco, I won't do that. In a sense, that experience made me a better and much more verbal leader. I know we can never do that again." What she means by "that" is letting go of her team. "I feel responsible for nineteen people losing their jobs in the UK."

For Angelique, the UK was personal. She'd put her entire life on hold for it. Her husband remained in Amsterdam and only saw her on weekends. She basically lived for work day and night and wanted nothing more than to have the UK work.

Though Matthijs was originally stunned by Invesco, over time his perspective shifted. There is no doubt in his mind that the client ended the contract without merit, but he no longer thinks it was all in vain as he initially did. The progress he and Angelique made in the UK market made an impression. Matthijs contends that their work may have "doubled, maybe tripled the valuation of the company." And he, too, learned valuable lessons from their time in the UK, albeit hard ones.

CHAPTER 16

WORK HARD, PLAY HARD— REWARDING THE TEAM

WHAT BUOYED THE TEAM during these ups and downs and all of the long days and nights at work was the camaraderie and support they found in each other. This wasn't an accident. From the beginning, Chris, Erik, Bas, Ilco, Lydia, Jan-Willem, Angelique, and Matthijs had each other's backs. They were leaders and believed it was their job to model this support for the whole Ohpen team. They were all underdogs in one way or another, and all believed in fighting to the end—giving their all and enjoying the ride. This "work hard, play hard" mentality was a driving and sustaining force for them, and for the success of Ohpen.

THE OHPEN WAY

From the beginning of Ohpen—even before Chris and Matthijs found the Rokin building and put in the full-service gym and boxing ring—sports and competition were a priority. It wasn't just how they blew off steam; it was also a healthy outlet for the intense and competitive natures of everyone on the team. Competition, in Chris's mind, goes hand in hand with ambition and the give-it-all mentality. Chris often plays Ping-Pong or challenges others in the boxing ring to see who has an internal drive. Jan-Willem remembers that in the very early days, when they were all practically living at the office, he had to bring several changes of clothes to work each day. Even "friendly" games of Ping-Pong with Chris would get so intense that by the end, Jan-Willem would be dripping with sweat. "Not one part of my clothing would be dry. That's how intensely we played," Jan-Willem says, laughing. The competitive nature is endemic throughout the company. It's in its DNA. Bas admits, "All

> Competition, in Chris's mind, goes hand in hand with ambition and the give-it-all mentality.

278

of us founders did sports at a high level. In Dutch there is a saying: 'If there's no friction, it won't shine.' In other words, we believe you have to be competitive to get the best out of each other. If we do something, we do it all the way. We don't want to lose. And we all really want to do well."

Though admittedly Chris would say hiring and developing the Ohpen culture in the early days came about organically and was largely based on intuition, in some ways it was also very intentional. From the beginning, Chris and the founders knew exactly what kind of culture they wanted to create and foster. Not one to suffer fools or laziness of any kind, Chris himself was on the lookout for exceptional human beings who were driven, who loved their work, and who weren't working just for a paycheck. Bas agrees and says when hiring that he is "always looking for people with an intrinsic drive. And that doesn't have to be work-related immediately; it can also be some kind of unusual hobby. That's where it all starts." He recalls hiring someone who played chess at a very high level, someone who played football at a high level, and another who'd started sailing around the world in his early twenties. Bas adds, "Almost everyone had something special about them. And if it's already in their work, that's even better. But drive and passion is where it all starts when building a culture. *The culture is the people.* If we don't give people work that is challenging, they'll eventually just walk away. If you have the right people, of course. You can have drinks every day and parties every week, even a bizarre salary, but good people will leave in the end if they don't like what they do."

Like Chris, Bas knows the landscape is changing in work environments and recognizes that most people spend more of their waking hours at work than they do at home, so the challenge is to create a company and culture that employees want to return to every day.

Though the company has grown and priorities have shifted over the years, one thing has held fast throughout it all—the Ohpen culture. Anyone who spends more than five minutes within the Ohpen building will be able to not only sense the "work hard, play hard" vibe; they'll actually hear these statements from any employee at any given time: "Give it all." "Exceed expectations." This is more than just lip service, though. It's palpable and it's *everywhere*—literally printed on large posters throughout the building. These statements are overheard in stand-ups (quick meetings before the day starts), in the kitchen area, and in informal meetings in their open-concept office areas. One day after an interview with a potential hire, an Ohpen recruiter and other Ohpen team members discussed the potential hire's merits:

"He has amazing credentials," one said.

"He's smart," another said.

"He looks great on paper," another said.

"*Too bad he doesn't fit the culture.*"

"Why?" someone else prodded.

"He didn't do his homework. He couldn't answer basic questions about what we do. He could have found that out just by looking at our website, some press releases, and other publications," one said.

"He didn't go the extra mile. Nothing exceeded my expectations," another added.

"He asked about what time he could leave each day and if he could work from home," someone said from behind their computer screen while attending to another task.

Everyone nodded their heads in agreement. *Not a fit.* It's not that they're judging others harshly or expect people not to have a life outside of Ohpen; it's just that everyone who works at Ohpen knows that *if you want to reap the rewards of working at Ohpen, you have to be willing to step inside the ring.* Of course, they're talking figuratively here, but they're also talking about the actual boxing ring in the fitness room in the basement. "Stepping into the ring is actually not a requirement," Matthijs says with a laugh. "The workout room is there because we value our employees' health and well-being. The healthier our employees are, the more productive they are."

For Chris, the ring serves another purpose. He can see who is willing to "give it all" (under controlled and safe conditions, of course, most of the time). Over the years, Chris has learned a lot about his employees by sparring with them in the ring or by watching them work out with the elite trainers he hires and makes available to the staff. Chris himself trains avidly in martial arts, practices yoga, and eats mindfully, and he encourages his employees to do the same, and not just for the obvious reason—that healthy employees are productive employees. Rather, Chris has always looked at what Ohpen does as almost Olympic-level competition. Working at Ohpen requires, he argues, the same level of stamina, endurance, strength, mental clarity, and passion to win that you need in order to compete and win a medal in the Olympics. Olympians train the whole body and mind. They exercise and watch their nutrition. Chris started reading about nutrition, looking into what elite athletes like boxers eat. He was also inspired by athletes, like tennis champion Novak Djokovic, who changed his diet and saw improved performance. Most notably, he observed the changes within himself when he changed his own diet. In addition to becoming more physically fit, he also became more mentally focused and sharp, and he wanted his employees to benefit from good health habits, exercise, and nutrition as well.

It's not always an easy topic to broach with employees, but Chris isn't one to shy away from difficult conversations. Chris recalls seeing a new developer smoking a lot. Like, really a lot. On top of the smoking, the employee was overweight and was calling in sick all the time. Chris couldn't help but notice and wanted to help. He approached the employee and said, "Hey, listen … you know, we have a gym downstairs. Maybe you should try it. We have personal trainers, too, and they can guide you and help you." The employee said that he would think about it. Then Chris told him that the following week they would have a conversation and that the employee could choose the topic. It would be about the gym session, or it would be about why he calls in sick so many times. Immediately, the employee was receptive and said he'd love to train.

In addition to personal training, Chris put the employee on a nutrition plan to help him lose weight. After six months on the plan, the employee wrote Chris a letter to thank him and told him that Chris's intervention changed his life. Chris recalls happily, "He told me he could finally pick up his kid, and he could never do that before!"

Of course, over the years some employees have been less receptive than others to Chris's advice. Chris recalls one particular employee who was so unhealthy and obese, he had difficulty walking. He was also addicted to fast food, Monster energy drinks, and smoking. So much so that he smoked two packs a day. "That's a lot of smoke breaks," Chris says. Chris was concerned and wanted to offer him help. He went to him one day and said, "Man, are you okay?"

The employee admitted he was decidedly not okay. "I don't feel very well. You know, I'm tired every evening."

Chris, not wanting to be impolite but also not one to hold back the truth, assessed the situation for his employee and addressed the situation directly.

"I don't want to be rude, but I think I know why you're tired. You're smoking two packs a day, and you're carrying around an extra hundred pounds. That has to be difficult. I understand, and I want to help you."

Chris offered him a personal trainer to join him in the gym. He also offered him nutrition advice.

The employee was resistant to Chris's help, but Chris was persistent.

"You're not even forty years old! I really want to help you, because you already have a lot of health problems."

While Chris was concerned about the man, he was also concerned about productivity and the accuracy of the employee's work, so he made his employee a deal. "I'll give you ten thousand euros right now for you to resign, because I think you're going to be sick a lot and then I'll have to pay for that, and I'll have to continue to pay you for two years. Or I can use the same ten thousand euros to make you better with a trainer and proper nutrition. I'll even pay for a nutritionist to help make your food. I'll personally pay for the personal trainer three times a week. I'm here in the morning, and you can join me during your working hours to train. I want to help you get better. So sleep on it, and then come back to me."

The next day, the employee came in and gave Chris his letter of resignation. Chris held up his end of the bargain and he paid his employee €10,000, and the man left and never returned. Chris admits the Ohpen culture is not for everyone—but for everyone at Ohpen, the culture is everything.

In fact, the entire company promotes this healthy lifestyle. Not only unhealthy employees are encouraged to participate in the group training sessions. *Everyone* is urged to participate. There are showers and places to change so they can work out any time of day. They can even leave their clothes there, and they will be washed for them. Same with the towels. Chris sometimes makes a joke, telling the employees that participation in the gym has a high impact on bonuses at the end of the year. After this "joke," more people always come and join the session. "I can be a bit intense sometimes, a bit overwhelming. I know it's not always right, but sometimes I just can't help myself," says Chris with a smile. On occasion, there are fresh cold-pressed juices served in the afternoons. Bowls of fruit are left on the counters—available for anyone. The kitchen, where employees are encouraged to eat together and gather, is not a standard company lunchroom where people bring prepacked meals and eat. It's a working kitchen. People make themselves "toasties" (a Dutch sandwich with melted ham and cheese), and some even cook full meals on the stove. On any given morning one might find an employee whipping herself up a frittata. In the evenings, employees who stay late to work are always provided with dinner. While some cook and eat in the kitchen, others play pool, Ping-Pong, or an intense game of two-person Tetris. Chris, Matthijs, and Angelique each have assistants who make them homemade nutritious lunches each day as they rush from meeting to strategic papers and back to meetings again. When guests or clients come, they, too, are offered these fresh, nutrient-rich, delicious office-made meals.

It's not just about being healthy and mindful, though, Chris admits. It's also about creating a completely different atmosphere for his employees and clients—a total experience that exceeds all of their expectations. "When clients or potential clients come in, it's a breath of fresh air. It's like, 'Whoa! This is different!'" Chris says. And that's exactly what he wants to be: entirely different—but not, as he says, "just for the sake of being different." For him it's much more intentional—and personal. "The idea behind it is this: if our employees are going to be at the office more than at home each week, why don't we make the office feel like home? In my mind, it has to be a place *where you want to be and where you like to be."* When he and Matthijs started looking for buildings in the early days, they looked at what his competitors were doing—setting up shop in massive, cheap, soul-crushing buildings on the outskirts of the city. "Who would want to go there? Show up there?" Chris wonders aloud. He knew if he didn't want to spend his days there, his employees wouldn't want to either. That is why they intentionally chose the center of Amsterdam. "We'll go right to the center of Amsterdam, because others are not. That alone will make us different. On top of that, I like walking to the office in the morning."

Every choice in creating and setting up the culture—right down to the design of the office—was intentional. "Even when we chose our building and set it up, we knew we wanted to be different. We didn't want these little cubicles just so we could fit in as many people as possible because it would be cheaper." In Chris's mind, "The building will be our marketing. When potential clients come in, they will see our building and they'll say, 'Whoa! I didn't expect that!' Now we've exceeded their expectations." Angelique considers the offices to be an extension of the Ohpen brand; it should also exceed expectations of the most valuable asset of the company—the team.

Though situated in a historic building, Ohpen's interior is quite modern and in many ways set up for a "homelike" atmosphere. There are closets to store coats for guests and employees. There are sitting areas on the rooftop to get some fresh air and soak up the sun (when it occasionally decides to make an appearance in Amsterdam). There are plenty of communal and other meeting areas inside as well. Murphy, a Welsh corgi, comes to work with his owner and greets most new strangers with a *welcome bark*, much like you would expect at any home, as people are allowed to bring their dogs and cats from the beginning. On Fridays, Ohpen provides pizzas and Heinekens and wine for every employee. No one is required to stay late on Friday to partake, but it is there for those employees who want to socialize. And often the founders and leaders are there too.

The kitchen area on Friday evenings looks a little bit like a meeting of the United Nations. There are people from Spain, Ireland, Russia, Ukraine, Bulgaria, the Netherlands, France, the US, El Salvador, and the UK. All told, there are twenty-nine different nationalities working at Ohpen. And this, too, is intentional. "I have different cultures in me, and I've always been interested in traveling," Chris says. "I've traveled to many countries, ever since I could afford it. I started backpacking when I was nineteen. The first time I went anywhere, I went to Asia. I saw all these different people, different cultures, and it made me realize that the world is *much bigger* than the Netherlands. We only have fifteen million people here; who are we to think that our way is right?" he asks rhetorically. This "open mindset" is part and parcel of the Ohpen culture today and crucial to the company's success. Different mindsets, approaches, thoughts, and ideas are welcome. "I have always liked learning from different nationalities, different people, different cultures, and I take a little bit from everywhere," Chris says. He adds that the beauty of the technology industry is that it "unites." He adds, "The product is the most important thing. We don't care if you're from a certain country, or you're a man or woman, or you're young, or you're old, or you're Muslim, or you're Christian, or you have pimples—all that stuff is not important. All that matters is that we create awesome technology together."

For Chris, creating an open, collaborative work environment with ambitious, passionate people who are united in purpose came out of a personal realization of what he and his fellow BinckBank employees had been missing while they toiled to build that bank from scratch. "When I worked at BinckBank, we worked so hard. Eat, sleep, work, eat, sleep, work for years, years, years, and years. And maybe on Friday, we would go get a beer at the pub next to the office, and then, maybe once a year, the company would pay for it, *and that was it.* There was always something missing." Chris, who values hard work, believes that those who work hard should be rewarded. When Chris rewards, he *really* rewards people. "My ultimate goal is to take people out of their comfort zone, and out of their zone in general." And Matthijs adds, "We look at rewarding employees in a different way than just throwing money at people. The financial industry is known for its golden cages, golden handcuffs. We don't want employees who won't move on because they are addicted to eating oysters, sitting in skyboxes in football stadiums, and waiting for bonus checks. We don't want people staying for the wrong reasons. We want people here because they're passionate, driven, and okay with an occasional reward—one that isn't guaranteed or even handed out every day."

> Chris, who values hard work, believes that those who work hard should be rewarded.

A DIFFERENT KIND OF REWARD

In the early years, one of the ways Chris rewarded his young team and took them out of their comfort zone was by taking them on a luxurious ski trip to France—complete with a chalet, housekeepers, and chefs—like the one he took the team on right after the team delivered Robeco's proof of concept. When Robeco finally did go live, again, Chris rewarded the team who worked on the platform and migration once again. Again, he rented a ski chalet, chef, flight tickets—the complete works. The trip, however, turned out to be a bit of a disaster for some—Erik broke his leg and required three surgeries, Andre cut his leg, James went temporarily blind (because he didn't know he needed to wear sunglasses), Rogier dislocated his shoulder, and Lydia impaled herself with a ski pole, bruising her and sending her to the hospital for observation. "We weren't hung over. We weren't drunk. We weren't going a hundred kilometers an hour. We were all just really stupid," Lydia says, laughing at the memory. "That was the last ski trip we ever went on," she adds with a grin. Erik chimes in and says, "It was just bad luck."

Erik was at the fun park, playing around on the ramps, when disaster struck for him. He did well—cleared five ramps—and he knew Bas was right behind him. So as he slowed and looked behind him, he hit a small bump he hadn't seen and fell backward. His leg literally snapped back over his ski boot, almost at a ninety-degree angle. "I was going maybe two kilometers an hour when it happened." Bas, an avid and accomplished skier, got to Erik right away, took off Erik's boot, and had to readjust the leg. Meanwhile, Erik says, "I was dying from the pain." Erik spent three of the days in the hospital, and when he finally returned to the chalet, it felt to him like "coming home." He was, in his mind, among friends, not colleagues. And they took care of him as such. "The bandages on my wound needed to be changed every day. It needed to be cleaned, and no one wanted to do it, but I couldn't do it myself." But one of his coworkers, who earned the nickname Medical Center that weekend, took it upon himself to take care of Erik. For Erik, Chris, and the entire team, this was not surprising—they took care of and looked out for each other. It's the Ohpen way. Though Erik couldn't do anything but lie on the couch for the remainder of the week, he still had fun joking around and bonding with the team when they came home each evening.

IBIZA

Another way Chris didn't so much reward his team but found ways to get them out of their comfort zones was when he would rent a house in Ibiza in the off season when there was no one on the island. He often invited all the IT guys, Erik, Bas, and Ilco, and a couple of the programmers to have their strategy meetings there instead of inside the office. "For a week, we would wake up to this every day," Chris says, pointing out toward his pool and sun-drenched garden with cascading pink bougainvillea along the walls of his Ibiza home. "We would talk about all our technological challenges and the future of our technology." But it wasn't all just business. "Work hard, play hard. We bonded. We got to know each other. We barbecued in the evening, made our own food, and talked." They talked about technology for sixteen hours a day. "But it didn't feel like work in an environment like this," Chris adds. "Here, everybody was relaxed. We actually got more done. We didn't have clients calling us; we didn't have office issues to deal with here. We would be here for just five days, then magic would happen. All of our best tech decisions usually came out of our meetings in Ibiza; all the architecture of our platform comes from those days in Ibiza," Chris says.

For Chris, the experiences away from the office are invaluable. Everyone's stress decreases, and they are able to think and see things more clearly. He doesn't mind spending money on experiences like this, either. Chris adds, "I think for the money we spend on reward trips or times away from the office, we reap the return times a thousand. On all these trips, our company was made better. We talked technology and actually made all the big technology decisions over the last ten years here or on the ski trip or doing the yearly AWS re:Invent cloud conference in Las Vegas. We use all our server-less technology because of the trip in 2017. We reprogrammed the platform to be able to scale better because of these trips. We have not been down in ten years and have the fastest system in the industry. We made all the important decisions when we were all together during these trips."

VEGAS

The trip to Vegas that Chris is referring to, where the company benefited from learning about new technologies, actually began as a "reward" or "thank-you" trip. In the summer of 2017, at the height of landing new clients, migrating new clients, and opening new offices, the team had more work than they could execute. They were nearing their deadline with de Volksbank, and they knew they weren't going to be able to meet the deadline at the rate they were working. Matthijs and Chris knew they would have to go to the Volksbank team and tell them, "We can't go live" if they didn't come up with an alternative solution and fast. Chris took aside the team working on the Volksbank project—around twenty employees—and said, "Listen, we cannot make it. We don't have enough time. So we're going to go to the client and say, 'Sorry, we're not going to go live. It's going to be two or three months later,' and I am the one that will have to tell that story. I have to 'sell' that. So I'll take the heat, and we'll do that. Or we could do this: We have three months—that's twelve weekends. And twelve weekends add up to twenty-four days. If you work twelve of those twenty-four days and work ten hours a day, it all adds up to the number of hours we need to finish this project. And if you finish, I'll send you all to the AWS cloud conference, re:Invent, in Vegas." Chris added a catch, though. He wanted the entire team to say yes. Everyone on the team had to be on board. He asked them to sincerely think about it. "It's the summer. We don't have a lot of summer days in Holland, so just say no, and I'm totally fine with it. And I'll go to the client and tell them, 'We're not going to make it.'"

Everyone but one agreed. However, the one who didn't say yes realized Ohpen wasn't the culture for him, and he was looking for something else anyway and quit—so the rest of the team could go. And the rest of the team then said, "Let's go for it."

In the end, the team pulled it off—and they all went to Vegas, the whole team, together with forty-five thousand other IT nerds, attending twenty-five hundred sessions, the whole week starting at 7:00 a.m. and finishing at 9:00 p.m. Angelique says, "In the end, the work hard, play hard mentality doesn't just benefit the employees; it benefits the clients."

It also benefits the company. Ilco, who self-proclaims, "I hate to fly, I hate jet lag, and I hate being away from home," went to Vegas begrudgingly. He isn't the type to need an incentive like a trip to get a job done. Nevertheless, he was genuinely curious about the Amazon re:Invent conference because, he says, "I can learn new stuff and see how they are doing things. I don't care about drinking or staying in a hotel together. I just want to focus on my job. That's what I like most." For Ilco, who always has his mind on programming, the event was eye opening. He got excited when he attended an event focused solely on AWS's serverless capabilities, Lambda. "With Lambda, which scales up automatically, we don't need to do anything on the infrastructure side of things. And I saw with my own eyes what other companies were doing and the enormous scale of those companies," Ilco says. For Ilco, the entire experience was rewarding, and in his mind the trip paid for itself. "I came home here and applied everything I learned." Ilco went to Chris with what he'd found out and recommended changes to the platform immediately. Ilco knew that in some ways it was a risk. Not to talk to Chris, though, who is always up for new ideas, but for clients. Ilco explains, "It is a risk for us as a company if we only focus on features for customers instead of improving our current code stack with new stuff."

Though he understands that it is logical to focus on clients and customers because they are who pay them to do the work, it's worth it, he says, "to do cool stuff to make our platform better."

Chris and Matthijs agreed, and in the end, Ilco says, "Now we're doing it. We are converting old stuff to the new upgraded AWS serverless—Lambda. We can now run code without provisioning and managing servers, and we only pay for compute time when we 'consume' it. So basically there is no charge when the code isn't running. And because of all that, we also do DevOps now. We have completely restructured ourselves to scale up."

What Ilco touches on here—the risk and reward factor—goes a long way. Some could argue it's a risk to spend on one's employees or to make sure their needs are taken care of, as there is a risk that employees can become "soft" or "entitled"—but in Chris's, Matthijs's, and the other founder's experiences, that hasn't been the case at all. In fact, the benefits of rewarding employees creatively and in very surprising ways has paid off in dividends. Thanks in part to this culture, the company continues to grow and is positioned to scale across the globe.

CHAPTER 17

SPREADING THE MESSAGE AND CIRCUMNAVIGATING THE GLOBE

FROM DAY ONE, THE GOAL AT OHPEN, though said with a wink of an eye, was "world domination." Matthijs admits it was a bit of a joke in the beginning between all the founders when they were just young guys. *We'll take over the world, mwa-ha-ha-ha.* In reality, however, the team has had a much more focused approach—building up the Netherlands office and team, establishing a solid culture that can go anywhere and that would thrive no matter what leader was in charge, and staying on message. Like everything else they do, the message they delivered would be very different than what banks had been used to hearing over the years. Chris credits this in part to Kalo at BinckBank. "Two of the best marketing approaches I learned were from him. I remember him saying, 'If I throw twenty tennis balls to you, you won't catch all of them, and I bet you won't even catch one. But if I throw you one, directly at you, you will.' Communication is about saying one thing and one thing consistently." Chris adds, "Building a brand is saying one thing. *Choose one thing, and do that.*"

That is why, over the years, *It. Just. Works.* has been a key message. If a bank only remembers one thing about Ohpen, Chris wants it to be that the platform works—all the time. It's reliable, adaptable, and compliant. *It will always work.* All the other stuff is there too—it doesn't go away, Chris says. "Yes, we operate in the cloud. Yes, we're secure. Yes, we are constantly redefining processes and have a completely different business model. Yes, banks don't pay us up front but rather pay a monthly fee. Yes, we have a different service level agreement. But I know that if I say all those things, I'll lose my audience. But *It. Just. Works.* they'll remember."

A DIFFERENT KIND OF MESSAGE

Another obstacle to getting the message out there is that Ohpen is a start-up, and though it is ten years old, it is relatively small compared to its other competitors. These major players have invested a lot of money over the years in raising their brand awareness, and there is no way a small start-up can compete on the financial investment plane. *So what is a start-up to do to compete with the likes of Oracle, Temenos, or IBM?* Chris says, "First of all, you shouldn't do the same thing. So you have to be *out there*," Chris argues. He also says people have to get comfortable with the fact that others won't like them. "So if we ask a hundred people what they think of us, we'd rather have five people say we're absolutely fantastic and ninety-five percent say we suck than everybody unanimously saying, '*You're okay.*' I knew we could never win against Oracle or IBM based on *trust*, on how *known the brand is*, on their *size*, or on *credibility*—all these marketing terms. We have to separate ourselves from the pack and be a different breed. Like *Blue Ocean Strategy*." In some ways being the underdog here has its advantages. Yes, the bigger companies have well-known and established brands, but that makes them comfortable—a bit too comfortable. And as underdogs, start-ups like Ohpen can afford to be creative, do things differently, and actually *be* agile instead of just saying they are agile. There is no risk of losing an established audience. As newcomers, Ohpen could afford to be witty, try new things, and find different areas to stand out—areas the bigger organizations never thought to engage in.

> And as underdogs, start-ups like Ohpen can afford to be creative, do things differently, and actually *be* agile instead of just saying they are agile.

Chris knew that every single aspect of the company and brand had to be different and stand out. The offices, as an example, had to be different than any other core banking company's offices. The website had to have a different feel and approach. For example, on the website, he wanted to showcase the talent of those who worked at Ohpen—and not as just static individuals with a list of degrees and accreditations next to their names and same old, same old pictures; he wanted banks to see that real people were behind the building of the platform and that they were all passionate individuals who don't only "give it all" at work but also in their lives outside, at home. Each employee has a page dedicated to them—what they like to do in their downtime and what motivates them to work so hard. At its core, the idea is that Ohpen is as good as its people, and its people are all extremely driven, self-motivated, and passionate individuals. "But," Chris is quick to add, "we don't want the genius a-hole. He can go work somewhere else."

HIRING AND RECRUITMENT AS A BRAND STRATEGY

Even hiring these employees is a different process than what other companies do. Yes, there are standard requirements and tests to see if candidates are qualified and can actually do what their résumés say they can do, but candidates also go through a rigorous interview process—not just with the HR recruiter but with various members of the team the potential hire will eventually work with. But other companies do that as well. Ohpen tries to select the energy they need in that team and in that moment. That is very abstract. It's a feeling. Every leader has their own particular style, too, as well as what they are looking for. Angelique will cut to the chase: "What sucks about you? If I call your best friend, your mom, your dad, your husband, your wife—what will they say?" And if someone responds superficially and tries to give her "something that they can turn positive," she knows they're not being genuine and honest. She also tries to see how adaptive they are; she'll change deadlines or move interviews. She knows if they become flustered, they won't last long. Being adaptive, agile, responsive, curious, and willing to do whatever it takes are attributes she is looking for in any employee. Though Chris also asks people the opposite of what they think is coming to keep people on their toes, in the end he believes *competence* is key. It's not just testing people for the sake of testing people. Everyone who works at Ohpen doesn't have to be witty or direct or creative or different. He explains, "When we are hiring in our finance department, I just want our books to be on time, and I want the books to be good. I don't care if they have a 'boring' personality. I don't care if they come in at nine and leave at five. Staying late doesn't make an employee 'Ohpen material.' It's if they are passionate about what they do, know it well, are able to get the job done, and execute it well."

Chris says that he and Matthijs always ask themselves, "Can he or she get stuff done?" when reviewing a candidate. "We need to be able to delegate. If we can't delegate, we can't grow," Chris explains. "Many companies have the opportunity to grow, but the leaders can't delegate, and then they don't understand why they are not growing. *Delegate.*" In addition to knowing they can trust someone enough to delegate to, Chris and Matthijs also look for a bit of grit combined with problem-solving skills. Chris recalls Kalo, his BinckBank boss, saying, "Everyone can be a captain on the boat when the weather is nice—when there is wind in the sails, there are no waves, and you have a good crew. But what happens if the waves are high, the wind is blowing too hard and in the wrong direction, and the crew is injured?" Most people don't like sailing during storms, Chris contends, so he likes to test people—in the ring, for example—to see how they will react when conditions aren't perfect. "Just asking them to do so will give me an idea."

Long before they hire, their recruitment strategy is a bit different as well. Even when attending recruitment tech conferences, they take a different approach. Chris noticed early on that he wouldn't be able to compete with the large tech companies in attracting talent at these types of events. Usually, the larger companies have massive booths and spend a lot of money on gimmicks to lure prospective hires to them. Chris knew there was no way Ohpen could compete if they just tried to replicate what the larger companies were doing. So their approach was completely counterintuitive. They would get the smallest booth space available—and they did not even make it into a typical "booth." Rather, Chris wanted just two computers, one big screen, and two Ohpen programmers to sit and work there with their backs to everyone. "Everyone else was greeting people and saying, 'Come on in!' And so I thought, 'We'll do the opposite. We'll have two programmers just sitting with their back to everybody.'" The reason for doing so, Chris explains, is that he knew programmers who were interested in code would be drawn to what was happening on the screen. Only those with a specific interest or passion would do so. Another reason for doing so was that he wanted to draw out the most curious individuals in the crowd—those that wouldn't be easily dazzled by the bells and whistles of massive booths but rather were looking for something more interesting and in line with what they were looking for in a career. He also admits not positioning someone at the front of the booth to welcome people in was out of empathy for programmers. "We work with a lot of programmers. Many don't feel at ease when marketing and recruitment people are coming at them. They're much more likely to engage with other programmers." Once the programmers were in the booth, they were invited to "play a game"—which was more like a test. When they moved the mouse to the left, the cursor would go to the right. "A good programmer would be able to fix this bug within sixty seconds. If you're mediocre, it might take several minutes. And if you take over ten minutes, then you can't program," Chris says. Another thing that was different was that Chris didn't want there to be any bull between the programmers and recruits. He wanted his programmers to be absolutely honest with them. "Every company has positives and negatives. I wanted potential hires to know the real deal and make the decision on their own. Open, transparent, and direct—right from the beginning." Chris's approach turned out to be a good one. "We got a couple of really talented programmers to join our company that day, and I think we had the cheapest budget in the whole place," Chris says proudly. "One of those hires is now the head programmer in the Barcelona office. He came from Iceland."

In an objective research study done by the job site Indeed.com, Ohpen was named as one of the top companies for attracting "top achievers" in the Netherlands. (Indeed's research was conducted online, and the respondent group consisted of 3,176 recognized top achievers and 824 unemployed job seekers and took place via the Decipher research platform of FocusVision.)

Top achievers, according to Indeed's research, were capable of strategic thinking and problem-solving. They were also passionate and had a genuine enthusiasm for their own task or role, and they had intrinsic drive that pushed them to go above and beyond. And finally, they had excellent communication abilities and could speak and write effectively with stakeholders.[8]

While they are proven excellent recruiters in the Netherlands, looking globally for talent is another thing Ohpen believes in. They're always looking for new ways to reach out to potential recruits and sometimes even using their own employees and events as a means to do so. When the entire development team went to Vegas as a reward for their hard work during the de Volksbank implementation, Angelique and her team—in particular, Carla Martinez—came up with an idea to capture the journey and give the team "*Mission: Impossible*: A mission should they choose to accept it." The team was asked to record themselves with GoPros during the event so that when they returned, they could edit and compile all the videos and make a recruitment video. The marketing team dressed all in black—*Mission: Impossible* style—and pitched the idea to the developers. The mission was accepted, and when all was said and done, they captured hundreds of hours of video of the team having a blast together in Vegas while attending the re:Invent conference.

Chris's advice to young start-ups is this: "If you're going to go, *you have to be different*. And if ninety-nine percent think, 'Look at them with their cheap-ass ugly stand,' that's fine. As long as one percent thinks, 'Hey, who are these guys?' you'll be okay." The reason he appreciates Angelique, who is now the chief commercial officer and manages the sales, business development, marketing, and PR team, is that she understands and shares Chris's desire to do things differently, but for a reason, not just for the sake of it. Chris commends Angelique's vision: "The whole Ohpen brand has been a cocreation between Angelique and me. We both loved building this brand, and it was funny that we were so passionate about it and so many developers did not really care about it. But just with the two of us, we created everything that is the Ohpen brand today. Those were some super cool things to do."

APPROACHING MARKETING AND PR IN DIFFERENT WAYS

Whether it's the building, the website, the talent, the recruitment, or even the RFPs, the Ohpen goal is ultimately to stand out, to "be a breath of fresh air," Chris says. Since they can't compete in size, they can do so in other ways. Most big companies, for example, have a massive writing pool whose only job is to answer RFPs. Often working from boilerplate, their responses are by rote, and since they're usually doing many at once, they're likely to hit the deadline without a minute to spare. At Ohpen, they do everything possible to "exceed the expectations"; that means that if an RFP is due on a Friday, they make sure it hits a potential client's in-box by Wednesday. And they make sure every answer to every question is, as Chris says, "a day at the beach." Chris adds, "We make every answer a piece of art. We try to exceed expectations there." As mentioned earlier, Ohpen spends nearly a thousand hours on an RFP, not just to make sure it's outstanding but so that they all fully understand the scope of the project. They take doing their homework very seriously. Nothing they do is just "getting phoned in."

They take the same personal approach when it comes to handling PR. They don't hand off journalists to a press officer. Chris, Matthijs, and Angelique personally respond to journalists' questions and go out of their way to develop personal relationships with journalists. In large companies, it's all about protecting the message and protecting the CEO. But because of Ohpen's size and mentality, they try to be as open, accessible, and transparent as can be.

REACHING THE RIGHT CLIENTS

Being *different* in the wide-open seas of fintech can have its challenges. Trying to build solid relationships with bankers isn't the easiest thing to do in a world where consistency, control, and legacy systems remain supreme. The whole point of building a brand, building a platform that works, creating a culture, and hiring the most talented programmers, of course, is to attract clients. Finding creative ways to engage bankers and potential clients is also something Ohpen strives to do differently. And in some ways, through trial and error over the past ten years, they have perfected it—in large part due to Angelique's out-of-the-box and off-the-cuff creative thinking. Once, while wrestling with questions like "How can I make control, security, and compliance sexy in an ecosystem where banks rely more on outsourcing to fintechs? How can I keep people's interest about this while at the same time assuring bankers that their own clients' data is safe?" Angelique had a spark of insight. She was on holiday in February of 2019 and was about to board her plane when she started thinking, "Whoa, this pilot *really* depends on a lot of outsourced parties to get this plane off the ground and keep it there. There is so much he depends on, and I as a passenger, a customer, depend on his being in control. In fact, my life, and the lives of the entire cabin and crew, depend on it, so if they are not in control—I actually feel it in my gut." She immediately concluded the following: *Banks are a lot like pilots: they depend on outsourced parties to build the machinery that they will ultimately operate and that their customers will depend on.* Her idea was to invite C-suite-level prospects as well as current clients to Schiphol Airport and let them try out a flight simulator—give them a feeling of not being in control in a new environment. She also arranged for a Royal Dutch Airlines pilot, someone from the Dutch Central Bank, and Bas to speak to the group about *being in control.* "We had clients talking to prospects, telling them how they viewed working with us. We had prospective clients having fun— quite literally 'enjoying the ride.' We had heads of bank boards of directors smiling and having fun—but the whole time they realized what the message was: *Ohpen is in control. They can trust us. It. Just. Works."* She also says the day worked on an entirely different level as well. By putting all these experienced, seasoned bankers together (who are really very accomplished at what they do) and asking them to be in charge of stuff they have *no* experience with gave them an appreciation for what Ohpen does. She adds, "We're the experts at this technical stuff. We know it *really well*, and we're in control. They can count on us." The day proved to be successful.

CIRCUMNAVIGATING THE GLOBE

"We do everything exceptionally well, with the hope of exceeding expectations and standing out," Chris says. When marketing, for example, it's common for large brands to sponsor sporting events. Oracle sponsors the Team USA sailboat, which costs the company millions of dollars. "There is no way Ohpen can do something on that scale, and most start-ups can't," Chris explains. Though not actively looking for sponsorships and places to put the Ohpen logo for the sake of having it there, Ohpen nevertheless ended up with the opportunity to sponsor a sailor—though not for the Volvo Ocean Race. Rather, in 2018 Mark Slats was in need of a sponsor for the Golden Globe Race right up until two weeks before he was to set sail. The Golden Globe Race is not for the faint of heart—or for someone without a sponsor or financial backing. It's arguably the most grueling round-the-globe solo sailing race in the world, because participants must travel the thirty thousand nautical miles *without* any assistance from the outside or modern technology. *So why would a technology company like Ohpen sponsor a sailor who wouldn't be able to use any technology while traveling the world?* The answer was simple: Mark Slats embodied everything the Ohpen team and brand were about—Mark Slats gave it his all. And, most importantly, he would give his all while conquering the great wide open.

Chris admits he also has a soft spot for underdogs. With only a week to go until he had to formally submit his entry, Mark had no sponsor and no money—only the sheer determination and belief in himself that he could do it. A mutual friend, Jeroen Swolfs, the photographer who captured *Streets of the World* (a compilation of one street scene from every country in the world), knew Mark and thought Chris could help him. Chris was busy the day Jeroen called him and said that he had a flight that afternoon, but if Mark could come by Ohpen's office in an hour, he would meet with him—it was the only time he had. Mark arrived immediately. Mark, known throughout the Netherlands as the Gentle Giant—a bulwark of a man over two meters tall with a long beard, gentle eyes, and a quiet, unassuming personality—proceeded to tell Chris his story. By trade he was a carpenter. He didn't go to school to be a sailor. He just wanted to do it because his mother was gravely ill with cancer. The sole purpose of racing for him was to raise awareness and money to help her fight the disease through an organization called Sailing-4Cancer. He further explained that he was up against seventeen other participants, most of whom had been preparing mentally, physically, operationally, and financially for years. In other words, he was the ultimate underdog. He'd even had to sell his own home, and now Ohpen was his last and only hope. Chris didn't need to hear much more. "I immediately saw the spitting image of our core values. I have never met a person that embodies 'give it all' as much as he does. The word *impossible* is not in his dictionary. Based on mental strength, tactics, and pure perseverance, he overcomes and takes on new challenges. Quite like the journey we have been on in building the world's first cloud-based core banking engine."

Through the sponsorship, Chris wanted to set an example for both his team and the Ohpen customers, namely "that we can achieve anything if we just set our minds to it." Within ten minutes of meeting and talking to Mark, Chris agreed to sponsor him and gave him the €100,000 he needed to officially enter the race. The only catch, of course, was that Ohpen's logo would go on the sails, and after the event, Chris wanted Mark to come back and talk about the journey at some events. After Mark agreed to the terms, the team hit the ground running. Chris turned the project over to Angelique and Carla. They handled all the details and even helped Mark stock the food he would need on the boat. It was an "all hands on deck" kind of moment for everyone involved, because there was no time to spare, especially as Ohpen was planning a rebranding that was due to be delivered when Mark was at sea. All of a sudden, a project that wasn't even in their purview now had to be delivered four months ahead of schedule as well.

On July 1, 2018, Mark Slats set sail from Les Sables-d'Olonne, France, and for 214 days he lived out at sea, faced seemingly insurmountable obstacles, grappled with unpredictable weather, faced massive storms that capsized his competitors' boats, fought for his own life when he was thrown from his boat during a storm, and cracked a rib when his lifeline catapulted him back onto the cockpit floor after first throwing him out to sea, hitting a toolbox on the way. Ultimately, the underdog, who had half the training and no sponsor until a week before entry, came in second place behind a Frenchman who avoided the storms altogether. Mark's harrowing journey was the ultimate symbol to the Ohpen team of the Power of One. He was alone, he solved the problems that came his way, he didn't give up, and he exceeded everyone's expectations.

Mark was the epitome of the Ohpen spirit, and on February 23, when he arrived in The Hague, several Ohpen team members, including Carla (who'd supported Mark's project throughout his journey and made all the arrangements) and Matthijs, were there to greet Mark when he arrived. As is tradition, several sailors and boats rode alongside him as he entered the harbor. People were lined up all along the sea wall, waving and cheering, to watch as Mark steered the *Ohpen Maverick* into port.

The team wanted to show that they could give it all as well, so Carla arranged sponsored clothing and video updates, took care of the banners, talked to the press, and organized press events—everything she could do to help Mark on his mission. Angelique adds, "It was really cool to see that everyone in the organization had this 'give it all' mentality."

In many ways, the "right place at the right time" coincidental nature of the meeting of Mark and Chris, the massive undertaking the race presented, and the fact that Mark himself was an underdog are the perfect metaphors for describing Ohpen and its dare-to-be-different mentality. Ohpen was formed by a bunch of underdogs who just happened to be together at the right place and the right time. They had a simple vision: they wanted to take the closed and archaic banking industry and make it open—transparent—and they wanted to take it around the world. They weren't going to let anything stop them. No matter what obstacles, storms, bad currents, attacks, injuries, threats, and changes occurred, they had a singular purpose and vision. Like Mark, who had to rely on his experience, talents, wits, knowledge, and sheer force of will to take his small sailboat around the globe, the Ohpen team did more with less. Like Mark, who couldn't falter, couldn't give up because there was no one else who could sail for him, the Ohpen team gave it their all. Like Mark, who traversed the globe—not stopping until his journey was complete—Ohpen won't either. Because unlike for Mark, the race isn't over; the journey hasn't ended. In so many ways, it's only just begun.

CHAPTER 18

THE FUTURE IS WIDE OHPEN

IN THE EARLY PART OF 2018, while everyone was in Vegas celebrating a job well done on de Volksbank, learning more about the industry, and taking notes about AWS's Lambda serverless capabilities, Chris stayed behind in Amsterdam. He was in the middle of yet another tense and stressful negotiation. A private equity firm wanted to buy Ohpen. In many ways, in just ten years' time, Chris had come full circle, and he found himself sitting in front of yet another investor.

Private equity firms from all over the globe had been calling Chris for many years. They all wanted a part of Ohpen. Investment bankers, corporate finance advisors, family offices, investment firms, competitors, and private equity firms all found a way to contact Chris wanting to meet up with the intention of buying shares in Ohpen. Chris always said no, but if he had to sell, then he would have preferred family-owned investment companies, because he believed they tended to have more of a long-term investment horizon than many private equity firms. "I always found it so strange that many of these PE firms' first and only questions related to (1) how much revenue we had today, (2) how much the following year, and (3) if we could go any faster. But they had to know that an entrepreneur finds more things important than just these numbers. I think sometimes you have to make decisions for the long-term strategy instead of short-term revenue, and many of these PE firms say they agree, but in reality they don't. They only want you to grow, but I am an entrepreneur for more reasons than just to grow," Chris explains. Another reason Chris preferred family-owned firms was that he knew the industry he was working in and that it changed slowly. "Our business is a marathon and not a sprint, so a firm that wanted a quick three-year ride was not going to find that at Ohpen." Chris also had personal reasons for how he chose the investor. "I wanted to be an entrepreneur, to be free, to feel free, so any investor that gives me the feeling that I will not be free anymore, I will not go into business with. If I have all the stress of being an entrepreneur and all this responsibility, then I also want the perks, and for me one of the most important perks is freedom. If I can't have that, then I might as well go work for a boss."

In some sense, the perfect storm was happening all over again. Interest rates were very low; financial markets were very high; PE firms had a lot of cash to invest; cloud and SaaS business started to take over different sectors; and another Dutch company, Adyen, went public and exceeded everyone's expectations. So when Michel and Chris talked, they both knew that this perfect storm might not happen again very soon. And because so many people wanted in, they could literally choose who to sell some shares to. The sale would mean that a new investor—in this case, the Dutch company, NPM—would essentially buy a 35 percent stake in the company. The main reason Amerborgh, the investor that had funded Ohpen from day one, wanted to sell part of their stake was that selling meant they could profit a great deal and finance ongoing and future arts and cultural products.

Just as a perfect storm had brought Amerborgh and Ohpen together, it was coming together again that would serve them both in monumental ways. "I met with the whole ecosystem of investing, and we could have made a bigger deal, a higher valuation, and so on. There was so much interest that it was a bit overwhelming, but where I am an emotional and spiritual person in my private life, I can be very strict and rational about business matters. More money is not the priority and does not impress me. So Michel and I talked with many potential investors from all over the world, and out of all the investors we could choose from, we wanted to make a deal with NPM because it's a family office like Amerborgh with a long-term view, only bigger. It's like adding a second, even bigger engine to our diving boat. Also, I clicked with Rutger, their managing director, and a personal connection for me is very important," Chris explains. Rutger Ruigrok, the managing director at NPM, made the decision to purchase the Ohpen shares largely because he was impressed by Ohpen's achievements, namely the successful development of a cloud-native core banking engine and the strong customer base Ohpen had created, not to mention the solid company and culture that was poised for future growth and expansion.

Though many people didn't know it at the time, Chris was battling ongoing health issues. He became sick more often, first with the flu, then bronchitis, and he started having trouble sleeping. For twenty years, Chris had been, in his own words, "going, going, going—working more than three hundred out of five hundred weekends" during his years at Ohpen. In the early years, he wouldn't leave the office until 10:00 p.m. He never had a vacation where he didn't work a minimum of four hours a day. He missed out on family events, birthdays, and life events. "I was never there; I was always working," he admits. Chris started to do some soul-searching. He needed a break. He had been working nonstop for twenty years, and the responsibility had started to feel heavy. "Your biggest strength is also your biggest weakness. I am full of energy and passion, and this energy can enlighten others, but it also cost me a lot of energy having to expend so much of it," Chris explains. He thought of taking a year off again, like he did after leaving BinckBank, but that would put the board in a precarious position. "It would be like saying, 'You watch the house for the year, and when I am ready, I'll come and take my job back.'" Chris knew that was unrealistic. It wouldn't set up a solid foundation for any leader who would be left in charge, and it wouldn't be fair to him either. However, there was no escaping it: something needed to change.

In 2016, he had already faced an existential crisis of sorts when he found a massive tumor on his back and had to have it removed right away because it was growing so rapidly. Though benign, it was a wake-up call nonetheless that *life was short and precious, and one's health couldn't be ignored.* And Chris knew his body was undeniably crying out for help. He was suffering from chronic sinus issues, headaches, and bouts of the flu. His immune system was so overtaxed that he couldn't fight things like the basic common cold. As someone who was avid about health and body conditioning, he knew this wasn't sustainable. One day during meditation, it was almost as if he heard this message: "If you continue to do this, everyone will be rich—but you'll be dead."

Meanwhile, the day-to-day staffing issues, client issues, and stressors that came with running the company weren't going anywhere. He was starting to feel that everywhere he looked, something fundamental was missing. "I mean, I don't want to say, 'Oh look, I did everything,' but the truth is that when you're a leader, everything comes to your desk. The conflicts with clients, the conflicts with staff—and so yeah, I was missing something in my life."

Chris continues, "The most difficult thing in our business is selling the software. That's the hardest thing. It feels impossible to change bankers' minds. But somebody has to do it. And I tried. And did. For ten years. I took so many planes, slept in so many hotels—I was always on the road for a presentation or a meeting." Chris was nearing complete burnout. Even before all this, while dealing with the tumor, Chris worried he might die unexpectedly or things might become more serious. He realized then that he needed to adjust his will. He was worried first and foremost about his wife, Myrthe. He wanted to make sure she was well provided for after all the sacrifices she had made for the company. "Whatever happened, she was with me there on that journey for all these years. When I came home, landing at 1:00 a.m., was home at 2:00 a.m., and left again in the morning at 7:00 a.m., she was there. And I wanted to take care of her so she would never have to worry about money again." He sold some of his shares to make sure she would be provided for, and that, he says, was the first step in letting go of the company—and, in many ways, the company letting go of its hold on him. He started to reprioritize. He also took a step back and noticed he didn't like the person he was becoming. He felt his old anger rising to the surface again, in a way it hadn't in years. The exhaustion was wearing on him.

Meanwhile, he was watching the company scale and grow exponentially before his eyes. In fact, at the time, they were in the middle of enormous growth and potential. At the height of its growth and with the company on the cusp of major expansion, Chris did what a lot of entrepreneurs lack the courage to do: he realized that today he was not the man for the job of scaling the company and taking it where it needed to go. He didn't want to fall into the trap that so many entrepreneurs and founders do: holding on too long, digging their heels in the sand, and believing that the company couldn't exist without them. Chris knew better. He also knew his team, and he trusted the foundation he had built. All the best companies, he knew, survive long after their founders have left. If the only thing holding a company together is the founder, then there is something fundamental lacking at the core of the company. Chris knew the products he and the team had created were outstanding. He knew the culture of Ohpen would survive beyond him, and in many ways his own leadership style and mentality were woven intricately into the DNA of his fellow founders and several protégés he had mentored over the years. In early 2018, Chris finalized his decision, and he planned his next move.

One that would surprise everyone.

Since all the important decisions that were ever made came out of Ibiza, in April 2018 Chris invited his friend Matthijs down for a visit. In many ways, it was the perfect setting. On the golf course near his home in Ibiza that overlooks the beautifully crystalline turquoise waters of the Mediterranean for as far as the eye can see, both men were standing on a new precipice together. This time, though, there were no proverbial manta rays they were going after. They had secured several elusive clients during their time together. There were no more dark depths to explore together; they had already seen each other on their darkest days. However, their zest for adventure, their mutual admiration, and their willingness to give it all and have each other's backs were all still there. They had been through so much together.

Out on the golf course in Ibiza, Chris turned to Matthijs and said, "I'm giving you the keys. I want you to run things. Be the CEO." Matthijs responded in the most Matthijs-like way possible. *He didn't say a thing.* Chris knew his friend well enough to understand what was going on. He knew Matthijs needed to let the news sink in. Matthijs showed no immediate emotions. Chris knew his friend well. He wasn't expecting a grand gesture or an outward display of enthusiasm—or even an immediate yes. He knew Matthijs took these matters seriously. Chris just continued to talk. He explained, "I know what you're going through. You've always been in the number two chair, the way I was the right-hand man for Kalo at BinckBank. And *I thought* I knew what being number one entailed. But it's totally different moving into the number one chair. You've always been number two, and now is your chance to be number one. But you'll never know if you're up for it and what it entails unless you do it, so I'll give you the keys and you'll know it then. That's the only way to find out."

Matthijs slept on it. The next morning, Matthijs woke up and said to Chris, without fanfare or enthusiasm but with a deep sense of understanding of the job that was being presented to him and the trust his friend was putting on him, "This is the next step for me. I'll do it."

Over the next year, the two laid out their plans. Chris would slowly ease out of his CEO role, eventually putting Matthijs at the helm alone to run Ohpen as the CEO. Angelique would be there as well. Chris believed wholeheartedly in Matthijs—in the entire board—who would lead the company in his stead. Chris says, "I think they should be very proud that we—me and the investors—didn't have to go looking outside for someone to lead. We never, ever considered going outside to find a CEO. That's the biggest praise you can get." This vote of confidence in Matthijs comes from not just twenty years of friendship and growth but from seeing Matthijs give it all, right alongside him.

On April 6, 2019, exactly ten years after starting the company on April 6, 2009, Chris officially handed over the keys, making Matthijs the new CEO of Ohpen.

It was an enormous moment for both men. Chris knew he'd built something that would outlast him. And Matthijs was ready to take the company to the next level. In so many ways, they were just scratching the surface of what was possible. In just ten short years, the small but mighty team had built a successful banking platform and put it in the cloud. They now had several clients, had plans to broaden their platform, and had expanded into other countries. The future was wide Ohpen.

OHPEN'S NEXT STEPS

BY JUNE 2019, CHRIS HAD MOVED TO IBIZA. The offices and desk that he and Matthijs had once shared on the fourth floor of the office on Rokin no longer had Chris's ocean books on the credenzas. Chris's side of the desk was empty—cleared of all his ephemera, Buddhas, and tikis. Matthijs was alone, sitting on his side of the desk. There was no one sitting across from him to spar with, to bounce ideas off in the middle of the day. He was the number one guy now. Streaks of white glistened around Matthijs's temples as he ran his hand through his hair. Besides the graying hair, there was something about his posture, his countenance, that looked every bit the part: he was a CEO now. He nodded and listened judiciously as his assistant ran something by him. He looked—one might even go so far as to say—relaxed. *He was ready for this.* He had been preparing for this moment for twenty years. And in many ways, he wasn't alone. He had Angelique serving on the board with him. He still had Michel at Amerborgh for support as well as his new investor. And Chris had agreed to stay on as an advisor and help during the transition. "Matthijs and I always do the big things together. Important client meetings, new contracts or contract renewals, and strategy. And this we will continue to do together for a bit. But I don't have anything to do with the day-to-day operation and running of the company," Chris says with a smile. Matthijs is up for the task. At the ten-year mark, Matthijs says, "We have an incredible foundation, and all the intelligence we now have allows us to move into other products or other countries much faster than we were able to previously."

The foundation they have, of course, is that they were the first and only core banking engine to be in the cloud. Though everyone is jumping on the bandwagon these days, this wasn't always the case, and it is something Chris is acutely aware of. He recalls being called a "stupid idiot" and being told time and again, "It's never going to happen." That didn't deter him or the team. He reflects on something Jeff Bezos said: "If you want to create something truly innovative, you need to accept being misunderstood for a long time." "We were programming a whole new bank, putting it in the cloud, and offering it as SaaS. We weren't the status quo," Chris says. "In the past three years, the world did a one-eighty on us. Now all the companies are saying they are in the cloud, and the only thing they can say is 'cloud, cloud, cloud, cloud, we have cloud.'" However, Chris argues, "Buyer beware." Not all that these vendors are saying is true. "There are people who now just say their old data centers are the cloud!" Chris says with emphasis. "They haven't changed anything!" Instead, he contends, they have changed their messaging to the following: "We're in the cloud too; *we have our own* cloud." But what they actually have is their old legacy system. And no one in the entire industry can compete with what Ohpen has in terms of experience. "You can't get this knowledge and this experience overnight. We've been doing it for ten years! We were first doing it, we were the only ones doing it, and we've learned a shitload. We've failed, we've failed *forward*, we've failed again, we've stood up again, and these experiences *no one else* has."

Chris admits that he is sure others can do this and others can learn this. "However," he adds, "everybody has the right to their own opinions, but no one has the right to their own facts. And the fact is, no one has been doing this longer than we have, because we were the first ones." This foundation Ohpen is operating with going forward to the future is so solid precisely because of this invaluable experience. Chris likens it to diving. He himself has done over five hundred dives. There is a part of the ocean off the coast in the deep south of the Maldives in which the current is so strong and treacherous that a person cannot dive without an underwater scooter. Chris argues it's worth the risk, however, because the area there is teeming with fish and wildlife and it is home to the best diving in the world. "But it is fucking scary if you don't have a minimum of a hundred dives," Chris says. "I would never, ever, ever say to anyone who has less than fifty dives to go dive there, because it's in the middle of the ocean, there's a shitload of current, there's a shitload of fish, and there are sharks everywhere. *You need experience or to be with someone very experienced.* "If you're a novice," Chris wonders aloud, "who do you want to go down in the heavy current with? A guy with five hundred dives or fifty? Or worse, none?" The same, he says, for going on the cloud. You need to go with someone with experience, who knows how to navigate the proverbial tough currents and the shark-infested waters *and* who knows what to do when the unexpected happens. Long story short, Chris says, "You need mileage. You need those hours, and it's the same with this. You can't just say, '*Hey, you're an experienced IT guy. Why don't you take my crappy old piece of software and yank it into the cloud so we'll also have a cloud banking system?*' It takes years and years of practice."

Confident in their experience and the foundation he and Chris built together over the years, Matthijs as the new CEO has several objectives and goals for the future of Ohpen. The first objective also happens to be a personal challenge—to service a Tier 1 bank. "My success as a CEO of this company will be measured by whether or not I succeed at meeting this objective. The proof, of course, will be when we are servicing millions of accounts for a single customer, all the while still maintaining that incredible *It. Just. Works.* promise." He is sure that Ohpen's platform and the talented team working on that platform can do just that.

His second objective is to take Ohpen to the pension market. This is a strategic decision that Matthijs and Chris agreed on even prior to Chris's handing over the keys. "The technology that we have created will be beneficial to the pension market. From an asset point of view, it is one of the largest markets in the world. Because in Europe at least, the majority of one's retirement capital is generated by the pension that one accumulates in one's professional life."

His third objective is to make sure Ohpen's products and services are available in all European countries. To take the platform international.

Matthijs's objectives go well beyond scaling the company. "From a product point of view, we have to make sure we stay relevant and that we stay abreast of the latest technological developments." His plan to do so requires that Ohpen invest at least 20 percent of their total resources into research and development. "We have built so carefully in the first ten years, and we have to make sure we are still innovative ten years from now." In his mind, that means using the cloud as optimally as possible and to only invest the time into the processes. "We don't want to build what Amazon already has," Chris adds. For example, they utilize AWS's Lambda serverless model and what Ilco's developers are doing all across the company. Matthijs adds, "Going serverless is just an example of the technological renewal and legacy killing that we're doing to make sure we don't pile up a lot of tech debt."

He has plans for the organization as well. While the company is well on the way to being a fully functional DevOps organization, they still have room for improvement. "We want to get rid of the idea that you have all these people with certain capabilities, which are all combined into departments. We want to make sure that instead of departments with software engineers or requirements engineers or support engineers, instead of organizing it like that, that we make autonomous teams based on a certain domain—DevOps." This, he believes, will help the company grow, but it won't necessarily mean needing to hire more staff. "With this current model, whether we add two to four new clients each year, we should feasibly be able to do it with the same team that we have today, give or take a couple of people." He believes this is even possible if they get a Tier 1 bank.

Matthijs explains, "We need to keep the organization operating in an agile way. We must make sure that we don't build this huge monster but rather an army of small superheroes that can run in opposite directions." Not one to be unrealistic, Matthijs is fully aware of the challenges he and the Ohpen team will face. "We have to keep our eyes open and not be a victim of the not-invented-here syndrome."

Chris adds jokingly with a laugh, "We actually invented that syndrome twenty years ago."

Matthijs doesn't believe in looking much further out than five years. In his mind, that is "old-world-economy" thinking. In this rapidly moving technological age, a company has to expect the unexpected, plan to unplan, and adapt to new and arising technologies and advancements all the time.

Matthijs believes he has a clear set of tangible and measurable objectives to achieve in the next three years. He wants to reach ten million accounts, including a Tier 1 bank, and establish Ohpen's platform as a "player of importance" in the pension investment administration. "If I can succeed in the coming years in making sure all these things happen, I will be a very happy camper."

Matthijs is well aware that having goals and executing them are two very different animals and that leading a company through a major leadership change won't be easy. But he adds, "Despite the changes, the vision and the mission of the company won't change with me standing at the helm. They stay the same. *It. Just. Works.*"

Matthijs is nothing if not deeply self-aware. He knows that from a management perspective, he brings something entirely different to the table than Chris. He knows he doesn't have Chris's charismatic or extroverted personality. But it's not personality alone that makes a leader, Matthijs contends. "It's all about giving the people perspective, inspiring them, and communicating well. Communication is key. It's explaining the objectives of the company and translating the perspectives. And I think communication is not only explaining and not only telling but also providing some context on why we made those decisions."

As the CEO, Matthijs knows his job is much more than providing people with a stable workplace and a monthly paycheck. It's also to inspire them and make them believe in something, that what they're doing contributes to something tangible and important to the world. "I think the work that we're doing here is important, and it creates value for the whole value chain," Matthijs says. "If we can demonstrate that we have our eyes on the future, that we put this dot on the horizon and say, *This is where we're going—come with us*, that we are consistent in this and finish what we started, we will do great things. People need perspective, inspiration, a vision, a clear direction, and a sense of purpose, and I think that I can offer that just as much as Chris, even if it is in a different way." That being said, Matthijs humbly says, "But I learned a lot from the way Chris led the company in the past ten years." Matthijs, though unassuming, adds with a bit of a sly look in his eye, "But I also think that there are things that he learned from me."

Matthijs knows it won't be easy for some. "There are certain people who thrived under Chris's leadership, but there will be certain people who will thrive more under mine."

Chris has total confidence in Matthijs. Again, he likens it to diving. "Matthijs is at the helm of the diving boat now. He decides where it goes. He decides who gets on the boat. He decides where the best opportunities for diving are. But it's still a diving boat. It's not a fisherman's boat. It's not a tourist boat. It's the diving boat that we built. We've been diving together into the depths for many, many years. The boat is still the same; it's just a little bit bigger, and he's the captain of that boat now—and what happens to that boat and on that boat." Chris is deeply aware that the culture Matthijs is inheriting on that boat is a personification of who Chris is. He knows that won't go away, but he also knows Matthijs is going to add his own personality to it.

Matthijs is confident at the helm and says that Chris was enormously supportive in helping him and the entire company with the transition. In April 2019, in true Chris fashion, he wanted to go out with a bang and celebrate and thank the team that had helped him build Ohpen. He rented out a gorgeous, as Matthijs calls it, "fairytale location" on the island of Mallorca. Chris threw an elaborate, no-holds-barred, three-day black-tie party, day trips (hiking, biking, and golfing), and spa visits for all of his guests. His assistant, Esther, planned everything right down to the finest detail—even a makeshift boxing ring for the DJ—and they shipped in the team's favorite Friday afternoon snack: fries from the local Amsterdam french fries vendor. No detail was overlooked. During the ceremonies, several Ohpen employees got up and spoke about the company and all the opportunities Chris had afforded them over the years. And in a symbolic gesture, Chris asked Matthijs to come forward. Then he literally passed him a blazing torch, making it clear to everyone in the company that *Matthijs was in charge now. Chris's work was done.*

Chris made it clear not just with that gesture but through meticulous planning prior to his leaving, ensuring that everyone in the company, and also all clients, knew that Matthijs was their go-to guy now. If there was an issue, they had to take it up with Matthijs. Though Chris promised he would always be there for Matthijs in an advisory role, he made sure everyone on the entire team knew that Matthijs was at the helm and steering the ship toward new horizons.

"I'm grateful he did that," Matthijs admits. And in the same breath, Matthijs also acknowledges that it couldn't have been easy. Though Matthijs himself hasn't been there yet—he can only imagine what it was like for his friend to turn over the keys to something he'd created first in his mind then actually executed the build of throughout his entire adult life.

Chris, however, admits that it was. "I don't miss it," he says with a smile from his home in sunny Ibiza.

However, once a month or so, Chris travels back to Amsterdam to check in. On a recent trip, the new receptionist stopped him at the door when he arrived and questioned him. "Who are you? Who do you want to see?" Instead of explaining that he was the one who had founded the place, Chris smiled and used his fingerprint to enter the building. "I'm here to see the CEO," he said.

After entering, he walks through the halls. Like Dickens's Ghost of Christmas Present, no one seems to see him or notice him. The morning stand-ups are in progress. The DevOps teams are preparing their work for the day. He walks upstairs, heads down the hall, and passes the conference room with the large four-foot-tall Buddha statue that he bought in Bali. Erik and some developers are meeting to discuss future products. He walks into the kitchen, where several people are gathered pouring coffees, whipping up breakfast, and checking in with each other. One grabs his gym bag and heads down to the basement to get in a prework workout. Chris heads upstairs again. Bas and Ilco are at the whiteboard, working their way through some flowcharts. Competing with each other. Ribbing each other with jokes. Making each other better. Chris smiles and moves on. He heads up another flight of stairs. Carla, Kees, and the marketing team are talking to each other from their desks, laughing about something Carla just said. Meanwhile, an HR recruiter has his headset on, conducting interviews for a new hire. He's trying to see if this candidate has what it takes to *give it all*. Chris passes Angelique, who is giving a tour of the building to a prospective client. She's working her magic. He can tell that the prospective client is impressed. Like they are inhaling a breath of fresh air. Just like he had always hoped they would. Then he passes Lydia's office. She's meeting with one of the five managers who report to her. The manager is older than she is. But she's got this, Chris thinks. "The young Padawan becomes a Jedi," he says, laughing to himself. Chris heads upstairs. He passes his assistants, busy making cappuccinos for Matthijs and a veteran client who is in the office and looking to expand an offering. Matthijs is commanding the boardroom across from the desk Chris used to occupy. *He's so got this*, Chris thinks.

The team of underdogs who everyone discounted, who everyone told, "It can't be done. You can't put a bank on the cloud," is still there. Doing it. Giving it their all. Even without Chris there. It may have started with Chris's idea. A dream. A dream of building a platform that *just works*. And together they worked tirelessly together to build that dream. And in the process, they grew up, too, becoming friends, brothers, and in some cases actual family. And though Chris is no longer at Ohpen to execute that dream, *the dream is still alive*. The people behind building that dream are still wholly committed to fully realizing it. And there is no stopping any of them. These underdogs, who everyone around them underestimated, all still have a lot left inside of them to prove. After ten years, they're still giving it their all and trying to exceed expectations on the big Ohpen diving boat carrying them toward that circular dot out on the horizon—the future. Ohpen's future.

APPENDICES

DEAR BANKER: HERE ARE TWENTY THINGS OHPEN WANTS YOU TO KNOW

1. Are You SaaS? Or SaaS ... ish?

Many vendors try to disguise old solutions and applications as SaaS by introducing a SaaS version of their legacy software. But why do you want SaaS? To make sure you can focus on your business and outsource the rest. Being truly SaaS means that it is scalable in terms of feature development and pricing. This automatically requires one version of the software for all clients, hosted in the cloud—like Ohpen's platform. If you're using anything else but true SaaS, prepare to get bills after every major change—that is, if you have survived the upgrade (because if it is not one version, it is simply a different business model for your vendor). So always check the following: Where do you host the software? Is there one version for all clients? Do they use cloud-native components or 'old stuff'? What does the delivery model look like?

2. You Don't Want Legacy? Well, Don't Migrate It, Then!

This is a hard one. We all want it. We all say it. And we all write it down as a core project principle. But in the end, hardly anyone has the courage to actually follow through. Why? Because if you do not migrate legacy, you have to say goodbye to some products, features, and sometimes even a few clients. Create a business case: What would it cost to maintain a legacy product (or data) on your new, flashy, steadily humming cloud-based core banking engine for the next ten years?

3. Red Pill or Blue Pill?

Simple. Stick to it. Dear board member: Please remember that the deeper into your organization, the more product- and process-loving people you'll have. This means that with the best of intentions, they will give it their all to have every little exceptional idea that pops into their mind included in the scope. Congratulations—you have just welcomed scope creep. If you want to implement within time and budget, set a razor-sharp and super-black-and-white scope for phase 1, phase 2, and so on. Everyone claims to work agile; few really adopt its principles. There are only two choices. You do it, or you don't. There is no middle ground here.

4. Rome Wasn't Built in One Day

When Steve Jobs and his team launched the first iPod, why did they not deliver an iPhone, as this was the end goal? Because it would have taken them a couple of years to do so. If you want to go fast and controlled, deliver an MVP and follow a phased implementation approach. Small iterations work well if you do them in a disciplined way.

5. Get Involved (Yes, You)—Must Have Senior (Board) Involvement

Prince2/waterfall, agile, and what other project methodologies have we tried without any real results when delivering large IT projects? If you want to stick to a strict scope and deliver it in a phased approach, you need senior—preferably board—involvement to keep project teams small, make fast decisions, and resolve issues. Set biweekly meetings of a maximum of sixty minutes, and make sure the project team only presents issues with decision options. We know that you already have too many meetings, but this time will pay itself back by reducing the implementation time by 50 percent (and therefore the costs).

6. Get Your Definition Straight: What Is Cloud?

Having a managed service by a data center company does not qualify as cloud. Many vendors label their solution as cloud, but it simply isn't. If you don't care, if you're only supporting a low number of customers and don't have to worry about too many peaks in your processes/requests, then don't worry about using the cloud. But if you do care and want a scalable solution that controls costs and is able to autoscale with you when you grow—then you will want the cloud.

7. What Does API Stand For?

You always want to be flexible, as your customers will change continuously. This means that easy integration and changing your proposition is key. We always advise you to check the granularity of the API calls and underlying processes they manage. Does the cloud banking vendor have a true API-first strategy? Will this really bring you a flexible solution, or do you need to fit in a straitjacket? Besides the functional part of the API, check if it can handle peak loads, what the latency is, if it gives you more (too much) than you asked for, if the documentation is of high quality, if it is backward compatible, and so on. In other words, what are your nonfunctional requirements?

8. Get the SaaS Playbook

Banks that are truly cloud based should be agile and innovative. Changes on the side of the bank should seamlessly work with the cloud provider. Goodbye legacy; hello agility. You should be able to adjust your proposition settings without requiring coding or IT (or your vendor). Check if there is a user interface or API/developer portal where you can make these adjustments. Ask a vendor simple questions: "I want to adjust my fee structure or add new investment products. Where can I do this?" If you get a blurry answer, run. This is your chance to escape future costs for changes.

9. SLAs Don't Lie

So check it yourself. Simple. When your teams have completed the RFP process and you have three top-notch vendors left, ask for anonymized service-level agreements (SLAs) and pick the one that best fits your requirements. And you can tell the quality of the service by the detail level of the SLA report. How are the calculations made that underlie the data in the report?

10. The Magic Words: Uptime and Performance

What keeps you up at night? Downtime and bad performance probably rank high in your "worst nightmare" list. Make sure your provider has good "hot failover" provisions and good disaster recovery/business continuity in place. With regard to performance, a well-executed load/performance test should give you the assurance that the provider can deal with X times your peak resource requirements. Ask for such reports!

11. Houston, We Have a Problem

Everybody will always state that they have never not met their SLA and have not been down before. Ask your cloud banking provider to give a compliance statement on these two items. Remember, downtime will happen, but it should be rare, and your vendor should be honest about it. Assurance reports should also provide clarity on how the provider deals with incidents.

12. If It Sounds Like a Duck …

Everyone has an agenda. And interests. And they will filter the information they give to you. If there is one kind of spreadsheet one can easily steer, it is a business case. So make sure all the angles have been thought of and that it is complete. And preferably have an independent person (like an auditor) provide you with assurance that the business case doesn't paint the wrong picture.

13. Apples and Oranges

SaaS pricing is different from on-premises pricing. In a SaaS model, about 68 percent of all costs come from the subscription fee (the rest from implementation, customization, and training). In an on-premises model, only 9 percent of the total costs originate in the software licenses. Make sure to add the following costs into your business case as well: customization and implementation, hardware, IT personnel, maintenance, training, application of fixes/patches/upgrades, downtime, performance tuning, rewrite customizations and integrations, upgrade-dependent applications, and maintenance/upgrade of hardware/network/security and database. In other words, don't compare apples with oranges.

14. Is Core Banking a Commodity?

There is no business value in small adaptations to core banking processes like corporate actions processing or interest rate calculations. Like electricity, it is a commodity, and banks should just plug into a centralized factory to process these activities. Real business value is created in the customer-facing front end or in the middleware and how the different parts of the bank's ecosystem work together.

15. Ask Not What Your Country Can Do for You; Ask What You Can Do for Your Country

Where do software implementations cost money, take time, and turn sour? The moment you ask the independent software vendor (ISV) to change their solution and make a bespoke one for you. Ask yourself what your company will gain by modifying the process in the software, and only ask the ISV to do it if you really must. It's way easier to change your processes to the software instead of the other way around.

16. Focus

Some vendors have three platforms that do the same thing, or they have a solution that can support a gazillion products but cannot cater to the needs of one product for the full 100 percent and the solution is delivered through SaaS or on premises. There is a big chance that it can do a lot but not one thing really well, and you end up with a mediocre service/product. If a company does SaaS, then they should only be doing SaaS. Only those companies can truly excel in SaaS.

17. Is the Provider Your Friend or Your Foe?

When you do these big IT projects, you need to be certain that the vendor will give it all. How have they managed their biggest risks? How did they mitigate keyman risk? How do you know they will always be there? How do they respond when the shit hits the fan or when you need to meet a deadline? Will they work weekends? In other words, will they engage in war as your ally or adversary?

18. Rent It

The whole cloud is about having on-demand access to computer resources. Owning those resources is a very bad idea. You lose value more quickly than you can gain benefits. Time and money spent on configuration and maintenance could be spent on activities that add value.

19. What Is Customer First?

Academic research has shown that successful companies are those that employ a relatively high number of people in customer-facing positions. Back office operations and IT staff (for nontechnology companies) should be minimized in favor of employees that engage with prospects and clients.

20. Standard Means Standard

Before engineering the wheel yourself, check thoroughly if someone has beat you to the punch. Use standard components available in the market rather than building everything yourself, at least if it is not a core component.

GLOSSARY OF KEY TERMS USED BY OHPEN

Audit, Risk, and Fraud Management Module contains all necessary features such as market abuse, unusual transactions, prevention of terrorism financing, antimoney laundering, and suspicious payments.

BPO means business process outsourcing. Ohpen is a BPO provider for banks who need to outsource their platform solutions.

Cloud Computing is a model for enabling ubiquitous, convenient, on-demand network access to a shared pool of configurable computing resources (e.g., networks, servers, storage, applications, and services) that can be rapidly provisioned and released with minimal management effort or service provider interaction.

Core Banking Engine is the behind-the-scenes engine that processes all deposits, payments, loans, most bank transactions, and customer data.

Customer Relationship Management (CRM) is a technology for managing all of a company's relationships and interactions with customers and potential customers.

CRM Module (also known as the Customer Relationship Management Module at Ohpen) improves efficiency in managing clients, leads, accounts, and contact history data. It also provides a visual dashboard to manage the workload of your teams and employees.

DDoS Attacks (Distributed Denial-of-Service Attacks) are malicious attempts to disrupt the normal traffic of a targeted server, service, or network by overwhelming the target or its surrounding infrastructure with a flood of internet traffic. DDoS attacks achieve their effectiveness by using multiple compromised computer systems as sources of attack traffic. Exploited machines can include computers and other networked resources such as IoT devices.

DevOps is a set of software development practices that combines software development (*Dev*) and information technology operations (*Ops*) to shorten the systems development life cycle while delivering features, fixes, and updates frequently in close alignment with business objectives.

Lambda is a service provided by AWS that lets a company run code without provisioning or managing servers. Companies pay only for the compute time they consume, and there is no charge when your code is not running.

Order and Fund Management Module at Ohpen is organized to maximize efficiency in order execution based on sophisticated aggregation algorithms, using multiple cutoffs and connections with the relevant transfer agents, exchanges, or other trading venues.

Payments Module enables you to process contributions and withdrawals safely to and from the savings and investment accounts. Ohpen supports both commercial banking payments and wholesale banking payments.

Reporting and Analytics Module at Ohpen includes numerous standardized and custom reports for all departmental needs, including operations, risk, client services, tax, and general reporting.

Robo Advice Module at Ohpen handles the configuration of suitability testing and the subsequent management of model portfolios that are linked to individual customer accounts based on risk profiles. Automatic rebalancing takes place based on life cycling, volatility, fund weightings, and many other variables.

Savings Accounts Module at Ohpen facilitates the administration of both variable interest rate accounts and fixed-term deposits, whether they are fiscal wrappers or not.

Serverless Computing is a cloud-computing execution model in which the cloud provider runs the server and dynamically manages the allocation of machine resources. Pricing is based on the actual amount of resources consumed by an application rather than on pre-purchased units of capacity. It can be a form of utility computing. Serverless computing can simplify the process of deploying code into production. Scaling, capacity planning, and maintenance operations may be hidden from the developer or operator.

Request for Proposal (RFP) is a document generated by a company or a bank to acquire outsourced services.

User Management at Ohpen handles the complete authorization and access management, controlling the segregation of tasks and responsibilities adhering to the "four-eyes principle." The settings section provides "customization of the platform through configuration," including fee schedules, interest rates, payments, and reports.

LIST AND BIOS OF KEY
OHPEN PLAYERS MENTIONED

CHRIS ZADEH

Founder and Chairman of the Board

April 6, 2009–Present

See Author Bios.

MATTHIJS ALER

CEO

July 1, 2013–Present

See Author Bios.

BAS WOUWENAAR

Cofounder and former CIO

April 8, 2009–February 20, 2020

Bas started his career at Getronics in 1998 and switched to BinckBank in 2000. As BinckBank was still in its start-up phase, Bas built up its IT division, where he was responsible for the entire IT infrastructure until 2004. Consequently, his responsibilities broadened to encompass team management of what was to become one of the biggest financial shared service centers in the Netherlands. In 2009, he joined Chris's mission as one of Ohpen's cofounders to develop the best core banking engine in the world.

Although they do not delight him, he is a connoisseur of engineering failures, and he loves reading about them. His interest is driven by the conviction that the better we understand the causes of failure, the more likely we can engineer systems that don't suffer from them.

ERIK DRIJKONINGEN

Cofounder and Product Chief
April 9, 2009–Present

Curiosity has always been one of Erik's main drivers—the desire to know how things work and why certain things are a success and others are not. This led to his first position at BinckBank, where, with the rise of the internet, two things first came together: the online world and finance. In this environment, Erik was able to learn firsthand from experts about the banking business, banking processes, IT, and more importantly how they all fit together. After a few years of being part of the business, Erik cofounded BinckBank's BPO business, where he had a focus on operational excellence, from both a process and an IT perspective.

After seven years of fun, learning, and hard work, Chris asked Erik to join Ohpen as a cofounder. With almost twenty years of professional experience in the online banking business, he is responsible for the core banking platform as chief product officer. Erik likes to ride his (racing) bike from the office to his home to his wife and two little girls.

ILCO VAN BOLHUIS

Cofounder and Lead Developer

April 8, 2009–Present

After studying computer science at the University of Amsterdam, Ilco started working at a small IT company. Two years later, it was time for a new challenge, and he found one at BinckBank. There, Ilco started by developing internal systems, and during the last two years, he mainly focused on BPO systems. Ilco joined Chris from day one as cofounder of Ohpen. He is involved in all aspects of the software development of Ohpen's core banking engine.

In his spare time, Ilco can be found behind a computer screen researching, learning, and developing cool stuff. A new important phase in his life started in 2014 when his first child came into the world. Besides spending time with his family, no matter what sport it is, Ilco likes to do it. Cycling, running, (wind)surfing, and skating are among his favorites.

LYDIA VAN DE VOORT

Former Chief of Staff and Board Member

June 22, 2009–TBC

As the first-ever employee of Ohpen back in 2009, Lydia has seen the organization grow from a small team focusing on developing the best core banking platform in the world to the international organization that it is today. After having managed various internal departments, including operations, services, marketing, and development, Lydia is now responsible for Ohpen's HR, facilities, legal, and finance departments as chief of staff.

When not at Ohpen, Lydia enjoys nothing more than spending her free time outdoors with her husband, her young daughters, and her family and friends. All of this ideally followed by a long table filled with food and wine and great music playing in the background.

ANGELIQUE SCHOUTEN

Board Member and Chief Commercial Officer

June 1, 2011–Present

Angelique Schouten has over a decade of experience with retail banking, insurance, asset management, and fintech companies. She was responsible for running the first direct-to-consumer robo-investor in the Netherlands. Angelique started her career at Ohpen as chief marketing officer, where she successfully led business development, sales, and marketing activities. In 2016, she established Ohpen's first international office and company in the UK and was appointed CEO of Ohpen UK. After this adventure, Angelique was promoted and joined the global board of Ohpen and is now responsible for the commercial organization. She holds a master's degree in business from Nyenrode University (where she met her husband, Eelco) and a bachelor's degree in economics and marketing management from Fontys University.

Angelique likes to spend her free time visiting sports matches, traveling, practicing yoga, and kickboxing, and she is a true foodie. Her motto is "Life is not about fitting in—it's about standing out." Twitter: @angelique4real

Angelique is coauthor of the book *Monkey Money Mind: How to Stop Monkeying Around with Your Money.*

JAN-WILLEM KOELEMIJ

Former Head of Barcelona
July 6, 2009–March 31, 2020

Jan-Willem's career started at Ohpen in 2009, just after the company was founded. As one of the first employees, he was part of the journey from start-up to scale-up. Over the years, he has been responsible for several engineering teams, implementations, and client relations.

In 2016, he was responsible for setting up the Ohpen Development Center in Barcelona.

He loves to spend time with his family and doing all kinds of sports, like soccer, running, cycling, swimming, tennis, and high-intensity training. When he was a student, he played soccer at a semi-professional level. For Jan-Willem, having positive energy is where it all starts.

MICHEL VROLIJK

First Investor, Amerborgh

Since 2009

Michel Vrolijk (1960) has been advisor to the executive board of Amerborgh International since 2020. From 2007 to 2019, he was a member of the executive board of Amerborgh. Since 1985 his positions have included director at Van Lanschot Bankiers, managing director at Merrill Lynch NV, and executive board member at ING Bank. He also served as chairman of the supervisory board at WestlandUtrecht Effectenbank.

OVERVIEW AND TIMELINE OF OHPEN

2000 Chris joins BinckBank

MARCH 2001 Chris and Matthijs meet at BinckBank

JANUARY 2003 BinckBank is awarded a banking license

2004 BinckBank goes public at €1.18

JULY 2007 BinckBank acquires Alex.nl

2008 Chris leaves BinckBank to travel the world

OCTOBER 2008 The stock market crashes

2008 Stroke of insight—Chris has an idea for a bank platform

2008 Chris pitches his big idea to friends Erik, Ilco, Bas, and Matthijs

2009 Chris secures funding from Amerborgh

APRIL 6, 2009 Chris registers the company

2009 The team moves into the Keizersgracht

2009 Lydia van de Voort and Jan-Willem join the company

2010 The team moves into the Herengracht

DECEMBER 2010 Ohpen purchases insurance company Robein

2011 Robeco sends out an RFP

SEPTEMBER 2011 Matthijs officially joins Chris and the team

NOVEMBER 2011 Accenture awards Ohpen the Accenture Innovation Award

FEBRUARY 2012 Ohpen signs agreement with Robeco, their first client

NOVEMBER 2012 Ohpen moves into the building on Rokin 111

2011–2013 The team works day and night to launch the platform

APRIL 2013 Robeco successfully migrates onto Ohpen's platform. *It. Just. Works.*

2013 Chris takes team on ski trip to celebrate successful migration

SUMMER 2013 Major software company offers to purchase Ohpen

SUMMER 2013 Chris and Matthijs close to closing deals with major UK clients

LATE 2013 Nationale-Nederlanden Bank (NN) begins talks with Ohpen

MAY 2014 NN signs contract with Ohpen (client number two)

MAY 2014 Jan-Willem takes over as project manager of NN implementation

APRIL 2015 Ohpen in talks with BNP Paribas

JUNE 2015 Ohpen signs Aegon Bank (client number three)

2015 de Volksbank announces its RFP; Matthijs takes the lead

MAY 2016 de Volksbank selects Ohpen as a provider (client number four)

JULY 2016 Angelique named CEO of UK Ohpen

SUMMER 2016 Jan-Willem moves to Barcelona to open Ohpen's first tech hub

DECEMBER 2016 de Volksbank signs contract with Ohpen

AUGUST 2017 Angelique Schouten appointed chief commercial officer

OCTOBER 2017 Ohpen wins UK Department for International Trade's Innovation Award

NOVEMBER 2017 Ohpen attends re:Invent conference in Las Vegas

FEBRUARY 2018 Ohpen raises €25 million Series C financing for international expansion

MARCH 2018 Lydia named CEO of Ohpen's UK office

JUNE 2018 Ohpen officially sponsors Mark Slats in the Golden Globe Race

JULY 2018 LeasePlan Bank selects Ohpen for both Dutch and German clients

OCTOBER 2018 Invesco pulls plug at the eleventh hour; team shuts down the UK office

APRIL 2019 NPM Capital acquires 35 percent stake in Ohpen

JANUARY 2019 Chris becomes Ohpen's chairman of the board; Matthijs named CEO

APRIL 2019 Ohpen celebrates its tenth anniversary; Chris passes the torch to Matthijs

OCTOBER 2019 Ohpen signs agreement with TKP and enters Dutch pension market

ACKNOWLEDGMENTS

CHRIS ZADEH

Where to start? When reading these acknowledgments at the end of a book, I am always wondering what I would do. Now that I am here, I understand how difficult this is. So before I start, I want to apologize to everyone that I do not mention here but who was part of Ohpen in any way, shape, or form. I want to thank everyone who was involved. Too many people to name or highlight, so I will try to keep it brief.

I want to specifically thank a number of people. First of all Mary, who helped us write the book. Without her, it would not even have been finished. I really enjoyed all our moments together thinking, talking, changing, working. It was really fun.

Second is Matthijs, my "everything" partner and the yin to my yang for the last twenty years. It was a pleasure and honor working side by side with you, and I'll probably miss you the most out of everyone. No one I have ever met—perhaps maybe only Tjade—has made me laugh as much as you have. Thanks for that.

Then of course Ilco, Erik, and Bas, my cofounders at Ohpen. We started this adventure the four of us in a basement in Amsterdam—and what an adventure it was. Blood, sweat, and tears can be tattooed on our bodies after the last ten years, but laughter, companionship, and intellectual challenges must be among them as well. Ilco, I can't understand how I ever lost a table tennis match against you, because you suck at it!

Alex and Michel, who trusted me with their money to start this journey—thank you. Everyone else at the time thought it was a stupid idea. "This will never work," is what everyone—advisors, colleagues, family, and friends—said to you when you told them you might invest in a company that wants to code a bank from scratch, put it in this cloud thing, and offer it to banks as a "factory" so they could outsource the administration of their bank accounts. Still, not listening to all the naysayers, you gave me a lot of money to start the company.

Esther, my PA, who was always there for me to support me and take things off my plate, so I could focus on the important priorities.

Leni Boeren, the COO of Robeco at the time of the signing of our first client, was the first to trust us. I will forever be thankful that she gave us a shot.

And last but not least, my wife, Myrthe, who always had my back, believed in me, and was there for me. Without her, I could never have pulled this off—ever.

MATTHIJS ALER

Much like the contents of our book, my acknowledgments are in chronological order. Therefore, my gratitude firstly goes out to my mom and dad (and their ancestors before them) for everything they taught me and for the convictions that they gave me, without necessarily making their convictions mine. The emotional DNA that transcends from your ancestors through your parents to you plays an important part in one's development. My DNA did not disable me from pursuing dreams, both in personal and in professional life, and they gave me some of the characteristics necessary to do what I do.

Secondly, I am very thankful to the Dutch schooling system. In elementary school and secondary school, I was taught the basics, and then, later on, Erasmus University provided me with a toolbox that allowed me to start working in a well-prepared manner.

Thirdly, a big shoutout to Thierry Schaap and Kalo Bagijn, the founders of BinckBank: my first job. A Dutch saying that is applied to people that start working for the first time is: "You get paid both in money and in experiences." The advice to be taken from this is that salary usually isn't the most important thing to look at, especially when you start your career. Maybe they used it as a phrase to pay below-market salaries, but it is undeniably so that the experiences gained as a rookie at BinckBank are irreplaceable.

The first person I met at BinckBank was Chris. One year my elder and quicker to start working, Chris helped me to become dry behind the ears during these first years at BinckBank. In awe of how he led some of the transformational IT projects at BinckBank, I was desperate to learn some of his tricks. During the past twenty years, he taught me some very valuable lessons, and in return, I tried to repay with the same favors in other domains. I guess we were pretty successful as a team and learnt from each other. Joining Ohpen while the adventure was already two years underway, he has always treated me—in every way—as if I were there on day one. Thank you!

Then in 2005, I met my future wife. Siena has been part of my every move since that moment. She took a leap of faith in 2006 when she quit her job in Amsterdam to join me to go to Paris where I launched Binck.fr. Siena always knows how to perfectly balance "support" with constructive and critical questions, therefore challenging me to make the best decision while looking in the mirror. Being CEO sometimes makes you feel lonely, but she always makes me feel that she is by my side. I also want to thank my children Sophie (2010) and Gijs (2011). Becoming a parent is the most rewarding life event possible. They made my life so much more rich and fulfilling.

Finally, I'd like to thank Mary for having the patience to write this book with Chris and myself. It would definitely have been another book if it weren't for her.

ABOUT THE AUTHORS

CHRIS ZADEH, FOUNDER, OHPEN.COM, #1 BEST-SELLING AUTHOR

In 2009, after fifteen years of working in the fintech industry, Chris Zadeh founded Ohpen, the first cloud-based core banking company in the world.

With a give-it-all mindset, Chris embarked on a journey to defy the odds that were stacked against him and exceeded everyone's expectations, including his own. While pursuing his dream to create the best core banking engine in the world, he was confronted with countless rejections—so many they almost broke him. But Chris never gave up.

After leading the company as founder and CEO for ten years, Chris now serves as Ohpen's chairman. He and his wife, Myrthe, live in Ibiza, Spain, with their five cats and their two dogs, Yucca and Myza. When Chris isn't traveling the world, looking for new ideas and inspiration, or shark diving, he is writing books, meditating, and practicing yoga and the martial arts. He is the author of the #1 Amazon best-selling book *Monkey Money Mind: How to Stop Monkeying Around with Your Money* (Forbes). Follow his adventures and what he's up to next on Instagram at @theyogiceo.

MATTHIJS ALER, CEO, OHPEN.COM

As CEO of Ohpen, Matthijs is leading the 150-strong team to achieve its mission of developing the best core banking engine in the world. In previous roles within Ohpen, Matthijs concluded agreements with major banks, led the implementation teams, migrated millions of savings and investment accounts, developed new propositions, and shared his views on cloud banking as speaker on leading conferences such as FundForum and the AWS Summit, among others. As CEO, Matthijs is focused on signing Tier 1 banks, successfully entering the rapidly changing pension market, and signing Dutch banks with an international footprint.

Before Ohpen, Matthijs was responsible for the launch of binck.fr and the rise of BinckBank in the ranks of the French online banking landscape. Matthijs studied at Erasmus University. Matthijs and his wife, Siena, live with their two kids, Sophie and Gijs, near Amsterdam. In his scarce free time, Matthijs likes to coach kids in sports and spend quality time with his family and friends.

ENDNOTES

1   Jane Gravelle, *Tax Havens: International Tax Avoidance and Evasion*. White Paper. https://fas.org/sgp/crs/misc/R40623.pdf.

2   ENISA. *Cloud Computing: Benefits, Risks, and Recommendations.* White Paper. pp. 9–11.

3   ENISA. p. 11.

4   https://aws.amazon.com

5   https://www.osborneclarke.com/insights/am-i-a-data-controller-or-a-data-processor-and-why-is-it-important-anyway/

6   https://aws.amazon.com

7   Sophia Yan, Charles Riley, and Matt Egan. "Brexit turmoil deepens: Dow down nearly 900 points in 2 days." June 27, 2016. Money.cnn.com. Web.

8   "Ohpen Scores Highly for Attracting 'Top Achieving' Talent." https://www.ohpen.com/ohpen-attracts-top-achievers/